RAND NATIONAL DEFENSE RESEARCH INSTITUTE

T0131060

Manpower Alternatives to Enhance Total Force Capabilities

Could New Forms of Reserve Service Help Alleviate Military Shortfalls?

Stephen Dalzell, Molly Dunigan, Phillip Carter, Katherine Costello,
Amy Grace Donohue, Brian Phillips, Michael Pollard,
Susan A. Resetar, Michael Shurkin

Prepared for the Office of the Secretary of Defense
Approved for public release; distribution unlimited

For more information on this publication, visit www.rand.org/t/RR3055

Library of Congress Cataloging-in-Publication Data is available for this publication.
ISBN: 978-1-9774-0294-3

Published by the RAND Corporation, Santa Monica, Calif.
© Copyright 2019 RAND Corporation
RAND® is a registered trademark.

Cover: U.S. Navy photo by Justin Oakes.

Support RAND
Make a tax-deductible charitable contribution at
www.rand.org/giving/contribute

www.rand.org

Preface

Developed in the first half of the twentieth century, the basic structure of participation in the U.S. military reserve components (RCs) still reflects many of the sociocultural trends of that era: RC members were more likely then to have a long-term civilian career with a single employer and established benefits, weekends free for nonwork activities, and two-parent families with only one parent working outside the home, to name a few. Changes in employment stability, family structure, and economic pressures since that time created both challenges and opportunities for how the RCs are used to meet national military requirements. Modifying assumptions about Reserve duty has the potential to improve RC member recruitment, performance, development, and retention in critical experience-reliant occupational fields—and it could stem projected manpower losses by providing alternative RC service options and enhancing recruitment of those in fields with highly competitive civilian industries, such as cyberspecialties, information technology, and aviation.

Against the backdrop of this context, the Office of the Secretary of Defense initially proposed this research when the U.S. Department of Defense was undergoing a review of alternative force mixes as directed by the National Defense Authorization Act (NDAA) for Fiscal Year (FY) 2017—including military technicians (MilTechs), civilians, and full-time support personnel—and the FY 2016 NDAA-mandated conversion of certain categories of MilTechs to civil service positions. In the course of that work, the Office of the Assistant Secretary of Defense for Manpower and Reserve Affairs concluded that it needed to explore broader programmatic improvements that would go beyond the MilTech program to enhance the manpower available for a variety of national requirements.

This research was sponsored by the Deputy Assistant Secretary of Defense for Reserve Integration and the Office of the Assistant Secretary of Defense for Manpower and Reserve Affairs and conducted within the Forces and Resources Policy Center of the RAND National Defense Research Institute, a federally funded research and development center sponsored by the Office of the Secretary of Defense, the Joint Staff, the Unified Combatant Commands, the Navy, the Marine Corps, the defense agencies, and the defense intelligence community.

For more information on the RAND Forces and Resources Policy Center, see www.rand.org/nsrd/ndri/centers/frp or contact the director (contact information is provided on the webpage).

Comments or questions should be addressed to the project leaders, Steve Dalzell at sdalzell@rand.org and Molly Dunigan at mdunigan@rand.org.

Contents

Figures

Tables

Summary

Developed in the first half of the twentieth century, the basic structure of participation in the U.S. military reserve components (RCs)[1] still reflects many of the sociocultural trends of that era: RC members were more likely to have a long-term civilian career with a single employer and established benefits, weekends free for nonwork activities, and two-parent families with only one parent working outside the home, to name a few. Changes in employment stability, family structure, and economic pressures since that time created both challenges and opportunities for how the RCs are used to meet national military requirements. Modifying assumptions about Reserve duty has the potential to improve RC member recruitment, performance, development, and retention in critical experience-reliant occupational fields—and it could stem projected manpower losses or provide alternative service options in fields with highly competitive civilian industries, such as cyber specialties, information technology, and aviation.

Study Objectives and Approach

The key to the successful adaptation of the RCs is in the answers to three questions:

---

[1] There are six RCs within the U.S. Department of Defense (DoD): the Army, Navy, Air Force, Marine Corps Reserves, and the Army and Air National Guard. A seventh RC, the Coast Guard Reserve, falls under the Department of Homeland Security. In this report, "RCs" will refer to the first six, unless a specific RC is named.

1.   What specialties—including new skill areas outside the existing military structure and culture—are most difficult to recruit and retain within the Total Force?

2.   Which parts of the potential military workforce could participate in some segment of the RC to a greater degree? These may be people whose civilian employment and lifestyles differ from the twentieth-century norms that govern RC structure.

3.   What policies are needed to connect those "unlikely reservists" to those unmet requirements?

The current RC system can still meet most of the requirements it receives from the services—usually through the use of extensive cross-leveling, or pulling people from several units. Thus, this series of questions points to a need to develop new forms of RC participation aimed at drawing on new sources of human capital to meet shifting requirements, not a revolutionary change in RC management.

In aiming to address the above questions, initial assumptions guiding the study were as follows:

- The military services face the greatest challenges in recruiting, training for, and retaining individuals in several key areas of expertise.
- Segments of the U.S. workforce possess some of these skills, but civilians are unlikely to serve in existing military programs because of the manner in which these programs are designed and administered. Some who currently serve in the RCs could do more while continuing to serve at less than the active component (AC) time commitment.
- Outside of the U.S. military, other countries, agencies, businesses, or nonprofit organizations may have developed other ways to access labor on a non–full-time basis, and some of these models or methods may be appropriate for use by the U.S. Department of Defense (DoD).

We analyzed the issue from both the demand and supply perspectives to explore and test these assumptions, employing an iterative, qualitative analytical process comprising an in-depth literature review, key stakeholder and informant interviews of 36 individuals across a variety of governmental and private-sector organizations, analysis of existing survey data of the U.S. population, and in-depth analysis of six case studies derived from a review of more than 40 potential comparative cases. The ultimate output of this exercise was the creation, through an iterative exercise across the research team, of a list of possible workforce constructs aimed at enhancing innovation in U.S. military personnel processes. The research team identified these workforce constructs through multiple, iterative explorations of the intersections between (1) the presence of an important hard-to-fill specialty, (2) the presence of an underrepresented population with a particular potential to fill that specialty, and (3) a concept for how to boost that population's participation and/or meet the labor demand, generally borrowed from the experience of another organization. Most of the research and recommendations presented here focus on all-new manpower for the RCs—individuals with no military affiliation. However, some apply to individuals who currently participate in an RC at the normal level (two days of training per month and two weeks of annual training) but could participate more.

U.S. Military Demand for Particular Specialties

There are two distinct ways in which alternative manpower sources could help meet service needs. First, new models could attract individuals who are currently unlikely to serve in any capacity. Second, there may be ways for new models to take currently serving individuals from the minimum level of RC service (approximately 38 days per year) to a much higher level—180 days or more. While the lists of specialties for the two sets of needs overlap (an overall shortage affects both the minimum-duty and extended-duty pools), they are clearly met by two different populations and require different changes to policies, regulations, or practices.

We reviewed recent testimony before the House and Senate Armed Services Committees, government reports, and previous research studies (both government commissioned and otherwise) to identify current and anticipated manpower shortfalls in each service, specialties for which it is difficult to recruit or retain personnel, and general trends in military personnel requirements. We then requested information from the services and conducted interviews with representatives from policy analysis organizations within each service. Of particular interest were details on current and anticipated personnel shortages in the RCs.

The data informing our study included quantitative measures of the varying degrees of shortfalls. However, because a lack of comprehensive data prevented us from measuring shortfalls in a standardized manner across the services, we focused on identifying shortfalls by specialty rather than measuring the degree of shortages or the amount of the gap that a given policy change might close. Future efforts to build on this research might seek to devise models that would enable an "apples-to-apples" comparison of shortfall data across the services or estimate the potential impact of particular policy changes on such shortfalls.

Categories of Shortages

We condensed many of the shortage specialties represented in more than one service into larger categories to ease analysis and interpretation. These larger categories represent the general characteristics of the associated specialties, although the generalization cannot fully account for the individual nuances of each specialty. In some cases, a category includes specialties that are not currently hard to fill, but those cases are the exception. The shortage categories we developed were as follows:

- cyber
  - offensive
  - defensive
  - computer/network technicians
- intelligence
- maintenance
- aviation
- medical professionals
- construction

- special operations forces
- explosive ordnance disposal (EOD)/chemical, biological, radiological, and nuclear (CBRN) specialists
- linguists
- chaplains
- transportation.

It is critical to note at the outset that not all shortages are created equal. For example, the military depends entirely on the private sector to produce nurses, chaplains, and lawyers. In other fields, services train personnel in particular occupations but may suffer when those trained individuals leave for better-paying civilian jobs. Appropriate responses to shortages would differ under each condition. To better understand these shortages, we sought to identify distinguishing characteristics of each shortage specialty type in terms of where and how personnel in a particular specialty tend to receive initial skill training and to compare how the specialty fares in the public sector versus the private sector. For instance, even when civilian education provides the original training and certification, the military typically has a subsequent program to refine the skills for a military environment—such as for turning a priest into a chaplain. Table S.1 shows how we coded such characteristics for each shortage specialty.

Potential Obstacles to Service

With so many military career fields showing systemic patterns of shortfalls across the services, it seems unlikely that the solution is simply to improve recruiting incrementally. The need for a more radical solution calls for an examination of new areas from which to recruit or an increase in RC participation among those who currently choose to serve, with a particular emphasis on how the RC population can more adequately supplement AC.

Table S.1
Demand and Sourcing Characteristics

Specialty Type	Private-Sector Shortage	Public-Sector Shortage	Low/No Private-Sector Demand	Private-Sector Trained[a]	Military Training Pipeline
Offensive cyber		X			X
Defensive cyber	X	X		X	X
Computer/ network technicians	X	X		X	X
Intelligence		X			X
Maintenance	X			X	X
Aviation	X	X			
Medical professionals	X	X		X[b]	
Construction	X			X	X
Special operations forces			X		X
EOD/CBRN					X
Linguists	X	X		X	X
Chaplains	X			X	
Transportation	X			X	X

[a] Here, we are focusing on the initial training in the general occupation. Even when civilian education provides the original certification, the military will have a training program to refine the skills, e.g., converting a priest into a chaplain, for instance.

[b] This category includes personnel specialties requiring military training only (such as medics in tactical units), civilian training only, and a mix of military and civilian training. Notably, the extent to which these individuals are trained in the private sector varies among the dozens of medical specialties. Those with a particular military nature, like medics in tactical units, are trained by the military. Those that are most like civilian careers, e.g., nurses, tend to be trained first as civilians. A third group has a mix of military and civilian training options.

Current Participation Requirements

To be eligible for RC service, individuals must first meet basic requirements in several broad categories.

First, per DoD regulations, there are detailed lists of personal history and physical characteristics that must be met (with some variation by branch of service). Applicants must not have any serious law violations or drug use or history of serious health problems. They must meet age, height, and weight standards; score sufficiently on an aptitude test (Armed Services Vocational Aptitude Battery); and pass a physical exam. However, many personal characteristic requirements that may otherwise preclude RC participation may be waived.

Second, after completing initial training (8 to 13 weeks) and technical training (the duration of which varies widely based on specialty), traditional reservists are required, at a minimum, to participate in reserve activities 1 weekend per month plus 2 weeks per year at a base that is a "reasonable commute" from where they live. Participation 1 weekend per month and 2 weeks per year requires that service members be physically present at their unit training site during these times. For some individuals, regular geographic proximity to their assigned training base may be a challenge. Another potential barrier to participation is the time involved. In this report, we conceptualize time constraints largely with respect to civilian work schedules and work-life balance.

Third, military personnel must be available for deployment. This should always be an assumption when it comes to military service, but this policy is accompanied by a mandate for corrective action. A February 14, 2018, memo issued by the Under Secretary of Defense for Personnel and Readiness states, "Service members who have been nondeployable for more than 12 consecutive months, for any reason, will be processed for administrative separation."[2]

---

[2]   Robert L. Wilke, Under Secretary of Defense for Personnel and Readiness, "DoD Retention Policy for Non-Deployable Service Members," memorandum, February 14, 2018. Note that there is a blanket exception for pregnant and postpartum members, as well as procedures for temporary waivers.

Barriers to Participation

One initial hypothesis of our research was that policies and regulations originally intended to give structure to RC participation might, in fact, amount to barriers that make participation by certain demographic or occupational groups less likely. We used three nationally representative data sources to help identify occupational categories with substantial potential barriers to service: the American Working Conditions Survey, the General Social Survey, and the American Community Survey. Descriptive results from the latter two surveys suggest (as hypothesized) that certain factors associated with work schedules—including irregular or on-call hours, split or rotating shifts, and frequent extra hours—may be especially hard to reconcile with RC service. Similarly, occupations with particularly taxing mental or physical requirements appear to be least conducive to service participation.

Alternative Workforce Constructs

Foreign Military Models

Alongside the United States, many countries maintain reserve forces to provide for their national defense. These RCs differ in size, composition, and relation to their respective country's AC. However, other countries' RCs may offer insights or suggestive models for consideration by the United States because they illustrate different ways to use reserve personnel to complement active forces, recruit RC personnel from new segments of the population, or structure reserve service to complement civilian life.

We selected four country case studies for exploration based on similarities to the U.S. military in terms of force structure, societal context, and operational employment, as well as the extent to which these other countries' RCs relied on models that may be relevant to the current problems faced by the U.S. military: the United Kingdom, Australia, France, and Estonia. Each case study provided at least one example of a reserve manning option that (1) was different in some way from current U.S. approaches and (2) targeted an identified shortage specialty or underutilized segment of the population.

U.S. Private-Sector and Non–Department of Defense Public-Sector Models

In addition to reviewing innovative models that foreign militaries use to access human capital, we identified models in use in other U.S. public organizations, including nondefense federal agencies, as well as in the private sector. Through a combination of interviews and a literature review of both primary and secondary sources, we determined how each model targeted a pool of potential reservists and offered terms of employment that expanded the pool of RC recruits while aligning with civilian employers' policies on work schedules and travel.

Definitive data on the share of workers in nonstandard arrangements are hard to come by, but there is a widespread perception that these arrangements are becoming more prevalent, a result of both workers and employers seeking greater flexibility. These examples are relevant to the RCs not only as examples of more flexible employment options but also as indicators of what today's working population expects (or can adapt to) and how technology can enable new alternatives.

Alternative Workforce Constructs for Innovative Human Resource Management

The final phase of the research involved identifying alternative workforce constructs for innovative human resource management, with a focus on the intersection of demand for specialties, untapped supply, and alternative models. We developed the workforce constructs through an iterative process, whereby the research team met multiple times to examine and reexamine our research findings and devise new RC-focused concepts to address the intersections of workforce demand, workforce supply, and alternative workforce models. Through these multiple iterations, we reviewed the list of workforce constructs to ensure that, as a group, they addressed as many critical specialties as possible, targeted the most clearly underrepresented populations, and leveraged the most promising models from other countries and other sectors. Table S.2 provides an overview of these workforce constructs and shows how they map to the shortage specialties.

Table S.2
Overview of Potential Workforce Constructs for RC Service

Workforce Construct	Description	Goal	Target RC Recruits	Potential Disadvantages	Shortage Areas Targeted
No Passport Required	Creates more lenient standards for RC service with regard to deployability outside the United States	Expand population of personnel eligible for RC service	Individuals otherwise able to perform RC duties but unable or unwilling to deploy outside the United States or to austere environments for extended periods	Not all military specialties would be suitable; radical departure from the status quo	• Cyber • Computer/network technicians • Medical professionals
Telereserves	Expands existing telework arrangements, allowing RC members to perform a broader set of tasks remotely	Break down location and scheduling barriers to reserve service	Individuals who have time to serve but face obstacles to reporting to a duty station	Security risks; technological challenges	• Cyber • Computer/network technicians • Intelligence • Medical professionals • Linguists
Reserves on Demand	Uses technology to increase schedule flexibility, with service opportunities that accommodate RC member's availability	Break down scheduling barriers to reserve service	Individuals with limited availability, such as business owners and single parents	May increase turnover and costs; may pose challenges to performance management	• Cyber • Computer/network technicians • Intelligence • Maintenance • Linguists • Transportation

Table S.2—Continued

Workforce Construct	Description	Goal	Target RC Recruits	Potential Disadvantages	Shortage Areas Targeted
Seasonal Worker, Seasonal Reserve	Maintains total service day requirement but spreads shorter service periods over more days	Accommodate blocks of time during which RC member is unavailable	Individuals in seasonal labor force or block-scheduling careers	Coordination challenges; changes to payroll and other systems; potential for skills to atrophy due to gaps in training	• Construction • Linguists • Transportation
Job Sharing	Increases amount of time served by allowing reservists to share duties, either performing the same tasks or discrete tasks that are part of the same job	Increase scheduling flexibility; facilitates more service by those who cannot accommodate full service demands	Currently serving personnel who wish to transfer from the Individual Ready Reserve	Coordination challenges; changes to payroll and other systems	• Aviation • Medical professionals
Part-Time Plus	Makes it easier for reservists to serve more than the minimum days	Increase scheduling flexibility while maintaining stability	Individuals with consistent schedules but greater-than-part-time availability, such as entrepreneurs, caregivers, and students	Supply of reservists may exceed demand	• Aviation • General unit support

Table S.2—Continued

Workforce Construct	Description	Goal	Target RC Recruits	Potential Disadvantages	Shortage Areas Targeted
Sponsored Reserve	Service contractor employees simultaneously work for their employers and serve as reservists	Ensure adequate training for military contractors	Service contractor employees	Potential conflicts regarding legal status of contractors accompanying the force, operational control (contractors' obligations to their employers), and long-term care and support for contractors who are injured while performing their duties	• Maintenance • Construction • Transportation
Wounded Warriors	Offers pathway to RC service for disabled veterans	Leverage the expertise of wounded warriors while minimizing competition with existing programs	Disabled veterans who are able to work part time but not currently eligible for RC service	Security risk and potential for abuse	• General unit support

Table S.2—Continued

Workforce Construct	Description	Goal	Target RC Recruits	Potential Disadvantages	Shortage Areas Targeted
Warrant Officer–Deacons	Creates warrant officer career path for chaplains	Expand pool of reservists in specialty with broad shortages	Individuals with an interest in religious ministry who are not interested in/able to seek ordination and commissioning as a military chaplain	Potential resistance from faith groups and targeted recruits; challenges differentiating between military and ecclesiastical duties	• Chaplains

NOTE: Special operations forces and EOD/CBRN would be difficult to fill through these alternative workforce constructs without significant policy changes. Thus, we were not able to map these specialties to particular alternatives.

Recommendations

A large and growing segment of the U.S. population is not a primary source of military manpower—not because of any objective deficiencies in their capabilities or patriotic spirit but because of conflicting obligations and constraints on availability. Meanwhile, current trends suggest that employment options that offer more advanced notice, stability, and guarantees will appeal to increasing numbers of Americans.

Significantly, this need not be a threat to military culture; indeed, interviews conducted over the course of this research also suggest that the idea of bringing people with previous military experience back into the RC workforce is gaining popularity. Moreover, the RCs offer an opportunity to experiment with different military work structures and parameters.

Finally, advances in technology have contributed to the development of innovative workforce models. Technology can enhance employers' ability to implement flexible scheduling practices; for example, analytics can identify workforce needs, and online and mobile technologies facilitate role sharing and shift swapping. Technology is also a major driver of telework and remote work options that can broaden the pool of talent from which employers can draw.

We recommend adopting one or more constructs for alternative workforce management in the RC environment. We have provisionally assigned them to three bins, in order of increasing barriers to experimentation in the near term:

1. Getting comfortable with technology: the Telereserves and Reserves on Demand programs would probably grow incrementally as new tools from the civilian sector become available and as military managers see advantages to adopting them.
2. Waiting for an advocate: The Warrant Officer–Deacon, Wounded Warrior, and Seasonal Worker, Seasonal Reservist concepts seem to have few institutional barriers. These workforce constructs could be implemented by a service with leadership support.

3. Changing culture, policy, and systems: No Passport Required, Job Sharing, Part-Time Plus, and Sponsored Reserves all change more fundamental terms of the RC experience. Establishing such programs would require a coalition of advocates to develop the personnel systems to manage a new category of reservists, revise policies and statutes, and gain buy-in from both AC and RC leadership. The programs' feasibility would depend on whether such changes added capacity to the Total Force and were viewed as worth the costs and effort.

Unless a senior DoD leader is ready to become a long-term advocate for one of the constructs in the third category, the most effective route to change would likely be one or more of the constructs in the second category, which could generate some momentum for broader changes.

Beyond recommending the adoption of specific workforce constructs, we recommend the following actions to continue developing the knowledge base on RC workforce challenges and potential solutions to adapt RC service to the current market for RC manpower:

- The Office of the Secretary of Defense (OSD) should continue to assess its access to required talent—now and in the future—as well as the extent to which current manpower policies enhance or reduce the propensity and ability of Americans to serve.
- OSD and the services should continue to assess the extent to which their workforce practices converge or diverge from common practices in the civilian workforce.
- The services should authorize their RCs to experiment with alternative work structures where a demonstrable need exists and where the alternative work structure appears likely to meet that need. To this end, the RCs should explore new service options that both reflect and complement developments in the civilian workforce.
- The RCs should regularly consider how technological innovation can promote greater innovation in when and where individuals

perform their service and the extent to which they need to be present at the same time in the same place to train successfully.

- Finally, OSD and the services should continue to support such efforts as duty-status reform, which will add more flexibility and simplicity to the system, and mirror advances in reserve force management adopted by allied countries, such as Australia.

Acknowledgments

We gratefully acknowledge the assistance of a number of individuals across DoD, the U.S. military services, foreign militaries, and the private sector who took the time to speak with us for this study. Although we cannot name them publicly, we are indebted to them for their assistance. We are also extremely grateful to both MG (retired) Gregory Schumacher and Jennifer Kavanagh for their extensive and helpful reviews of this report. At RAND Corporation, we thank John Winkler and Lisa Harrington for their management support, Craig Bond, Sarah Meadows, and Saci Detamore for their management of the quality assurance process for this report, and Lauren Skrabala for her writing and editorial assistance. The Office of the Deputy Assistant Secretary of Defense for Reserve Integration deserves special thanks for funding this research, and we would especially like to thank COL John Moreth and CPT Eric Johnson for the guidance that they and their team provided throughout the course of this study.

Abbreviations

AC	active component
ACS	American Community Survey
ADF	Australian Defence Force
AGR	Active Guard and Reserve
ALP	American Life Panel
AWCS	American Working Conditions Survey
BLS	Bureau of Labor Statistics
CBRN	chemical, biological, radiological, and nuclear
CCDCOE	Cooperative Cyber Defence Center of Estonia
CDU	Cyber Defense Unit
CNA	Center for Naval Analyses
CS	chronically short
CWS	Contingent Worker Survey
DoD	U.S. Department of Defense
DoDI	U.S. Department of Defense Instruction
DSMT	Dual-Status Military Technician
EDF	Estonian Defense Forces
EDL	Estonian Defense League
EOD	explosive ordnance disposal
ESR	engagement à servir dans la reserve (engagement to serve in the reserve)
FAIR	Federal Activities Inventory Reform Act
FEFCWA	Federal Employees Flexible and Compressed Work Schedules Act
FMLA	Family and Medical Leave Act

FTS	full-time support
GAO	U.S. Government Accountability Office
GSS	General Social Survey
HR	human resources
IG	inherently governmental
IMPA-HR	International Public Management Association for Human Resources
IRR	Individual Ready Reserve
MilTech	military technician
MOS	military occupational specialty
NDAA	National Defense Authorization Act
NIFC	National Interagency Firefighting Center
NORC	National Opinion Research Center
NSF	National Science Foundation
ODA	Office of Disaster Assistance
OPM	Office of Personnel Management
OSD	Office of the Secretary of Defense
PMF	Presidential Management Fellows
QWL	Quality of Working Life
RC	reserve component
ROWE	results-only work environment
SDSR	Strategic Defence and Security Review
SERCAT	Service Category
SERVOP	Service Operation
SHRM	Society for Human Resource Management
SSS	Selective Service System
TWM	Total Workforce Model
USFS	U.S. Forest Service
UTA	unit training assembly
UTC	Unit Type Code
VA	U.S. Department of Veterans Affairs
WIAS	Worldwide Individual Augmentation System
WO-D	Warrant Office Program for "Deacons"

CHAPTER ONE

Introduction

The basic structure of participation in the U.S. military reserve components (RCs)[1] still reflects many of the sociocultural realities of the twentieth century, when it was developed: RC members were more likely then to have a long-term civilian career with a single employer and established benefits, weekends free for nonwork activities, and two-parent families with only one parent working outside the home, to name a few. Changes in employment stability, family structure, and economic pressures illustrate the extent to which the context of RC service has evolved since that time, and it will likely continue to do so. These changes may create both challenges and opportunities for the use of the RCs to meet national military requirements.

Parallel to these extensive shifts in U.S. society and the labor market that are described in detail in Chapter Four, the past two decades have accelerated the U.S. military's shift toward maintaining an operational reserve. In 2013, the Reserve Forces Policy Board offered the following definition of this new dimension to the RCs:

> Routine, recurring utilization of the Reserve Components as a fully integrated part of the operational force that is planned and programmed by the Services. As such, the "Operational Reserve" is that Reserve Component structure which is made ready and

---

[1] There are six RCs within the U.S. Department of Defense (DoD): the Army, Navy, Air Force, Marine Corps Reserves, and the Army and Air National Guard. A seventh RC, the Coast Guard Reserve, falls under the Department of Homeland Security. In this report, "RCs" will refer to the first six, unless a specific RC is named.

available to operate across the continuum of military missions, performing strategic and operational roles, in peacetime, in wartime, and in support of civil authorities. The Services organize, man, train, equip, resource, and use their Reserve Components to support mission requirements following the same standards as their active components. Each Service's force generation plan prepares both units and individuals to participate in missions, across the range of military operations, in a cyclical manner that provides predictability for Service Members, their Families, their Employers, and for the Services and Combatant Commands.[2]

This evolution in the RCs shapes the demand signal for RC manpower in several ways. On one hand, the increased level of readiness required places a premium on regular individual participation in some kind of training or sustainment. Therefore, alternative manning solutions will have to overcome concerns that participants are less ready than those attending traditional drills and annual training. At the same time, sustaining an operational reserve increases the quantitative demand for committed reservists and guardsmen, and if the population is not meeting that demand under traditional models, DoD must look for ways to access and retain ready participants under alternative programs.

Against this backdrop, the Office of the Secretary of Defense (OSD) initially proposed this research when DoD was undergoing a review of alternative force mixes as directed by the National Defense Authorization Act (NDAA) for Fiscal Year (FY) 2017—including military technicians (MilTechs), civilians, and full-time support (FTS) personnel—and the FY 2016 NDAA-mandated conversion of certain categories of MilTechs to civil service positions. In the course of that work, the Office of the Assistant Secretary of Defense for Manpower and Reserve Affairs concluded that it needed to explore broader programmatic improvements that would go beyond the MilTech program

---

[2] Arnold L. Punaro, Chair, Reserve Forces Policy Board, "Report of the Reserve Forces Policy Board on the 'Operational Reserve' and Inclusion of the Reserve Components in Key Department of Defense (DoD) Processes," memorandum to the Secretary of Defense, January 14, 2013a, pp. 1–2.

to enhance the manpower available for a variety of national requirements. Modifying assumptions about Reserve duty has the potential to improve RC member recruitment, performance, development, and retention in critical experience-reliant occupational fields. Additionally, modifying assumptions about Reserve duty could stem projected manpower losses or provide alternative service options in fields with highly competitive civilian industries, such as cybersecurity, information technology (IT), and aviation.

The premise of this study, therefore, is that answers to the following key questions will enable the successful adaptation of the RCs:

1. What specialties[3] are most difficult to recruit and retain within the Total Force? These may be new skill areas, outside the existing military structure and culture.
2. What parts of the potential military workforce could participate in some segment of the RC to a greater degree? These may be people whose civilian employment and lifestyles are also different from those twentieth-century norms.
3. What policies are needed to connect those "unlikely reservists" to those unmet requirements? Because the current RC system still meets most of the requirements it receives from the services, this primarily suggests an evolution to develop new forms of RC participation aimed at bringing in new sources of human capital to meet shifting requirements, not a revolutionary change in RC management.[4]

---

[3] Each service has its own term for a person's designated specialty, the skill set required for a given position, and the career-management model that connects the two: Military Occupational Specialty (MOS) in the Army and Marine Corps, Air Force Specialty Code, and Rating in the Navy. Unless we are referring to a specific service, we will use the term *specialty* to describe all of these throughout this report.

[4] While this is true, it is worth recognizing here that in many cases, requirements are met through extensive cross-leveling and predictable deployment schedules. A pre-9/11 RAND Corporation report defined *cross-leveling* as "[moving] soldiers from one unit to another to ensure that each has enough qualified soldiers for the required jobs"; now it is often done sequentially, with a donor unit giving up personnel to make a deploying unit more complete, and then becoming the recipient unit when its turn to deploy approaches. It is therefore likely that if the services required multiples of the same unit types simultaneously, the RC would

Indeed, modifying assumptions about Reserve duty has the potential to improve RC member recruitment, performance, development, and retention in critical experience-reliant occupational fields. Additionally, modifying assumptions about Reserve duty could stem projected manpower losses or provide alternative service options in highly competitive civilian industry marketplaces in fields such as cyber/information management and aviation.

Approach

The study team's rationale in approaching the research was based on the following assumptions:

- The military services' challenges in recruiting, training for, and/ or retaining individuals are not equally spread across all specialties—some specialties face inherent challenges in this regard.
- Segments of the U.S. workforce possess some of these skills, but U.S. civilians skilled in these areas are unlikely to serve in existing military programs because of the manner in which these programs are designed and administered. There are others who currently serve in the RCs but could do more, while still serving at less than the active component (AC) time commitment.
- Outside the U.S. military, other countries, agencies, businesses, or nonprofit organizations may have developed other ways to access labor on a non–full-time basis, and some of these models or methods may be appropriate for DoD.

To explore and test these assumptions, we employed an iterative, qualitative analytical process comprising an in-depth literature review, key stakeholder and informant interviews of 36 individuals across a variety

---

quite possibly fall short of meeting requirements. Nonetheless, it is not within this study's scope to outline a revolutionary shift in RC management. For the earlier usage and analysis, see Bruce R. Orvis, Herb Shukiar, Laurie L. McDonald, Michael G. Mattock, M. Rebecca Kilburn, and Michael G. Shanley, *Ensuring Personnel Readiness in the Army Reserve Components*, Santa Monica, Calif.: RAND Corporation, MR-659-A, 1996, p. 1.

of governmental and private-sector organizations,[5] analysis of existing survey data of the U.S. population, and in-depth analysis of 6 case studies derived from a review of more than 40 potential comparative cases.

The ultimate output of this exercise was the creation, through an iterative exercise across the research team, of a list of possible workforce constructs aimed at enhancing innovation in U.S. military personnel processes. The research team identified these workforce constructs through multiple, iterative explorations of the intersections between (1) the presence of an important, hard-to-fill specialty, (2) the presence of an underrepresented population with a particular potential to fill that specialty, and (3) a concept for how to boost that population's participation and/or meet the labor demand, generally borrowed from the experience of another organization. The methods employed in this exercise and its findings are discussed in greater detail in Chapter Seven. Note that these workforce constructs are not intended to provide a feasible solution to meet every demand signal; nor is there one for every source of additional manpower. Nonetheless, the list as compiled should be an effective guide for future innovation and additional research.

The results of this study are presented in the chapters that follow. Chapter Two reviews current DoD use of the RC and civilian personnel to augment its regular forces. Chapter Three defines the challenges facing each service in associated with accession and retention of personnel in specific skill areas. Chapter Four reverses the perspective and documents six segments of the U.S. workforce that are currently not leveraged to meet military requirements and presents a concep-

---

[5]   These interviews are attributed anonymously throughout this report in compliance with the U.S. Federal Policy for the Protection of Human Subjects (also known as the Common Rule). Both RAND's Institutional Review Board and human-subjects protection reviewers in DoD approved of this research method for this study. Organizational affiliation is included in the citation for each anonymous interviewee to give a sense of the individual's background and experience, but it should be noted that interviewees were not asked to represent their organizations in a confidential way. While interviewees were asked to respond based on their professional experiences, they were, in all cases, speaking for themselves rather than for their organizations in an official capacity.

tual model of key limiting factors across occupations. Chapter Five explores case studies in which other countries have utilized RC systems to meet their needs, and Chapter Six looks at private and public entities that have used innovative intermittent, part-time, temporary, shared, or other employment models. Chapter Seven combines the analysis of labor demand, supply, and processes to develop and explore alternative frameworks for human resource management within a broad RC construct. Chapter Eight discusses the implications of this analysis and offers recommendations for potential new paradigms for RC personnel management.

Most of the research and our recommendations focus on all-new manpower for the RCs: individuals who currently have no military affiliation. However, some of the recommendations apply to individuals who currently participate in an RC at the normal level (two days of training a month and two weeks of annual training) but could participate more. Points that are especially applicable to this population are highlighted throughout this report.

Relevance and Applicability of This Study

The findings in this report will be relevant and applicable primarily to U.S. military leaders and U.S. government policymakers. It is intended to provide specific recommendations and more general analysis that will be immediately applicable in developing and administering programs to facilitate access to civilian human capital to help meet DoD's most pressing requirements.

This research will also be applicable to broader U.S. academic and policymaking communities that seek to understand how organizations are adapting to changes in the workforce and labor market.

To the degree that other countries face similar challenges in finding personnel to meet military requirements, international audiences may also find the analysis and recommendations useful in guiding their own policy initiatives.

CHAPTER TWO

Policy and Practice Surrounding Current Reserve Component Personnel Systems and Related Sources of Human Capital

To establish the contextual background against which options for change are to be developed, this chapter reviews the current models, programs, and systems for accessing human capital in a part-time or temporary form. The RC provides extensive force multiplying options to the AC force of 487,500 individuals in the Army, 186,100 individuals in the Air Force, 335,400 individuals in the Navy, and 329,100 individuals in the Marine Corps.[1] Indeed, as delineated in greater detail below, the entire U.S. RC—spanning the Selected Reserve, Individual Ready Reserve, and Inactive National Guard across all services plus the U.S. Coast Guard—totals over 1 million individuals. The RC is not the only force-multiplying option for U.S. defense forces. Indeed, both civilians and contractors now provide force-multiplying capabilities, with the U.S. Department of Defense Total Force now conceptualized to include the more than 700,000 civilians currently employed by DoD, as well as DoD-hired contractors, in addition to U.S. service members. Relevant policy and practice related to each personnel option is discussed in further detail below.

---

[1] Government Publication Office, "John S. McCain National Defense Authorization Act for Fiscal Year 2019," January 3, 2018, p. 99.

Current Reserve Component Options

Although we do not review all the various statuses and statutes pertaining to the RCs here, we do provide an overview with a focus on the general models that guide the creation of these programs, because the remainder of our analysis will, in many ways, attempt to imagine new models that can replace or supplement the current ones and open the door to new programs and opportunities.

Thus, 10 U.S.C. sets the stage for the discussion, stating,

> The purpose of each reserve component is to provide trained units and qualified persons available for active duty in the armed forces, in time of war or national emergency, and at such other times as the national security may require, to fill the needs of the armed forces whenever more units and persons are needed than are in the regular components.[2]

The heart of these RCs is called the Selected Reserve, those programs where personnel are actively managed for the potential contributions to national security.[3] A broader category, the Ready Reserve, includes the Selected Reserve and other individuals with prior military service but less active participation, generally lower readiness, and a higher bar for them to be involuntarily mobilized. An even broader category is all mobilization-eligible individuals, which primarily adds eligible retirees to the count. Table 2.1 shows the way the various types and groupings of the RCs are organized, along with the total figures for DoD as of January 2019.

Traditional Reservists

Although each service has its own formal and informal names for it, the dominant paradigm for most RCs for decades has been participation in a formal unit, along with monthly training and a longer annual training event. When needed, these reservists are activated and may

---

[2]   10 U.S.C. 10102, Purpose of Reserve Components, January 12, 2018.

[3]   10 U.S.C. 10143, Ready Reserve: Selected Reserve, January 12, 2018.

Table 2.1
Categories of Reserve Service

Total DoD Reserve Mobilization Potential (3,133,244)					
Ready Reserve (1,026,896)				Standby Reserve	Retired (AC and RC)
Selected Reserve (793,058)			Individual Ready Reserve/ Inactive National Guard		
Paid Drill Strength (Traditional Reservists)	Active Guard/ Reserve	Individual Mobilization Augmentees			
700,505	79,976	12,577	233,838	11,153	2,095,195

SOURCE: Office of the Secretary of Defense, "DoD Total Military Strength: January FY2019," provided to authors by sponsor.

serve with their peacetime unit or another organization. These personnel are referred to as drilling reservists (U.S. Navy Reserve), traditional reservists (U.S. Air Force Reserve), M-day personnel (for "mobilization day," Army National Guard and Air National Guard), or troop program unit soldiers (U.S. Army Reserve).[4] Under 10 U.S.C., these individuals are required to "(1) participate in at least 48 scheduled drills or training periods during each year and serve on active duty for training of not less than 14 days (exclusive of travel time) during each year; or (2) serve on active duty for training not more than 30 days during each year."[5]

Each of the services organizes and utilizes its reserve components differently. By design, the Army Reserve and Army National Guard organize primarily into combat, combat support, and combat ser-

---

[4]   U.S. Navy, "Navy Reservist Roles and Responsibilities," webpage, undated; U.S. Air Force Reserve, "About," webpage, undated; U.S. Army Reserve, "Ways to Serve: Troop Program Units (TPUs)," webpage, undated.

[5]   10 U.S.C. 10147, Ready Reserve: Training Requirements, January 12, 2018. The exceptions include personnel in the delayed entry program and other specific programs.

vice support units that can deploy as part of, or alongside, traditional active formations. Outside of its unit structure, the Army Reserve also includes the Army's Individual Ready Reserve and retirees available for mobilization, and also contributes individual augmentees to the active Army or other requirements (see following). The Navy Reserve organizes itself into units that provide staff augmentation, personnel replacement, or other functional support to the active force, as part of either active or reserve units. The Air Force Reserve and Air National Guard are organized into flying units and support units similar to those of the active Air Force, as well as into support cells and head-quarters organizations capable of deploying or being tapped to augment active units. The Marine Corps Reserve is organized into tactical units that resembles the active Marine Corps; however, Marine Corps Reserve units typically deploy at the battalion or company level, often as additional force structure for active-duty units. Notably, the Army National Guard and Air National Guard also fulfill state responsibilities when not serving in a federal status and may mobilize in a state status as well under the command of their respective governors.

Individual Mobilization Augmentees

A second part of the Selected Reserve are individuals assigned to AC units, not RC ones. DoD policy describes individual mobilization augmentees as follows:

> trained individuals pre-assigned to an AC or a Selective Service System (SSS) billet that must be filled to support mobilization (pre- and post-mobilization) requirements, contingency operations, operations other than war, or other specialized or technical requirements.[6]

Individual mobilization augmentees are required to perform annual training, typically with the AC unit, familiarizing themselves with the unit procedures and becoming a known quantity ready for

---

[6] DoDI, 1215.06, *Uniform Reserve, Training, and Retirement Categories for the Reserve Components*, Washington, D.C., incorporating change 1, May 19, 2015, p. 25.

activation and duty with that unit. Depending on the specific program, they may also serve up to 48 training periods during the year, like a traditional reserve member, but this may be done on weekdays throughout the month to complement the AC and civilian work schedule.

Active Guard/Reserve Programs

Separate from the use of RC personnel to meet operational requirements, 10 U.S.C. and DoD Instruction (DoDI) 1205.18's recognize a requirement for the RCs to "maintain a cadre of FTS personnel who are primarily responsible for assisting in the organization, administration, recruitment, instruction, training, maintenance, and supply support" of those components. This function, usually labeled FTS, is primarily done through two programs. First, 10 U.S.C. authorizes each service secretary to call personnel to active duty "to perform Active Guard and Reserve duty organizing, administering, recruiting, instructing, or training the reserve components."[7] Current DoD guidance is that Active Guard/Reserve programs "will be administered as career programs that may lead to a military retirement after attaining the required years of active federal service,"[8] as opposed to the less lucrative retirement benefits most qualifying National Guard and reserve members receive when they reach age 60. While the primary purpose of the Active Guard/Reserve programs is to support RC personnel and units, this does not preclude program participants from supporting the full range of DoD operations, and many have deployed overseas either as part of their assigned RC units or as individual augmentees to other kinds of units.

Dual-Status Military Technicians

Dual-Status Military Technicians (DSMTs) are the second major type of RC FTS. Following the gradual elimination of non-dual-status military technician (purely civilian employee) authorizations through the

---

[7] 10 U.S.C. 12310, Reserves: For Organizing, Administering, etc., Reserve Components, January 12, 2018.

[8] DoDI 1205.18, *Full-Time Support to the Reserve Components*, Washington, D.C., May 12, 2014, p. 8.

2017–2019 National Defense Authorization Acts, all current "Mil-Tech" positions are for Dual-Status Military Technicians (DSMTs)." A DSMT is a federal civilian employee who—

(A) is employed under section 3101 of title 5 or section 709(b) of title 32;
(B) is required as a condition of that employment to maintain membership in the Selected Reserve; and
(C) is assigned to a civilian position as a technician in the organizing, administering, instructing, or training of the Selected Reserve or in the maintenance and repair of supplies or equipment issued to the Selected Reserve or the armed forces.[9]

Neither the Navy nor Marine Corps utilize the MilTech program.

Individual Ready Reserve

The standard U.S. military enlistment contract runs for a total of eight years, regardless of the length of the initial active-duty or Selected Reserve obligation, with the remainder to be served somewhere in the Ready Reserve. Those who do not join an RC unit or other Selected Reserve program generally go into the Individual Ready Reserve (IRR), creating a pool of recently separated veterans at junior levels who can be used as individual augmentees as may be necessary. Because members of the IRR are outside the Selected Reserve, they have no requirement for training in statute[10] but may be required to muster periodically to verify they can be mobilized if needed. They may volunteer for training or be activated for extended training or mission support. The services have also implemented, at times, programs where members of the IRR voluntarily do additional duty, earning points for retirement if not pay. They may also be specially managed as a way to cultivate particular skills, as in the Navy's Strategic Sealift Officer Program, discussed in Chapter Seven due to its similarity to sponsored reserve pro-

---

[9]   10 U.S.C. 10216, Military Technicians (Dual Status), January 12, 2018.

[10]   10 U.S.C. 10144, Ready Reserve: Individual Ready Reserve, January 12, 2018.

grams. They may also be mobilized involuntarily under higher levels of mobilization.

Summary of Current Reserve Component Options

As shown in Table 2.2, the total number of personnel available through any of these programs adds up to more than 1 million individuals.

Reserve Component Training and Force Development: Implications for Reserve Component Service and Structure

Reserve Component Training

The military RCs generally provide training at the team or local level. The military components, in particular, have extensively developed training programs, including individually developed or exercised skills (e.g., marksmanship, airmanship); formal training or education for full-

Table 2.2
Strength of the Reserve Components (as of January 2019)

Component	Selected Reserve	Individual Ready Reserve and Inactive National Guard	Total Ready Reserve
Army National Guard	330,992	1,650	332,642
Army Reserve	189,821	98,089	287,910
Navy Reserve	58,267	42,703	100,970
Marine Corps Reserve	38,324	62,957	101,281
Air National Guard	107,074	0	107,074
Air Force Reserve	68,580	28,439	97,019
Total	**793,058**	**233,838**	**1,026,896**

SOURCE: Office of the Secretary of Defense, "DoD Total Military Strength: January FY2019," provided to authors by sponsor.

NOTE: Selective Reserve includes the Troop Program Unit, Individual Mobilization Augmentee, and Active Guard/Reserve programs.

time active duty personnel; unit-based sustainment training specific to unit missions; and participation in national-level exercises. Traditional reservists serve for 39 days a year, traditionally allocated between 48 drill periods taking place over approximately 12 weekends per year, as well as two weeks of annual training.[11] Certain reserve units participate in more training based on the type of work they do (such as aviation units), their required level of readiness, or their plans for future deployment. Individual reserve personnel in these units may also participate in more than the 39 days of training each year if they are pursuing individual training opportunities.

Developing the Reserve Component Force: Monetary Benefits and Pay

Participating RC members receive one day's active-duty base pay for each unit training assembly (UTA; commonly two pay periods per calendar day) and one day of active-duty base pay for each day on annual training or other extended duty. While they may not receive full benefits for shorter periods of duty, when mobilized or serving in an Active Guard/Reserve program, they generally receive the same benefits as a regular service member, such as a housing allowance and paid moves between duty stations. They also are eligible for enlistment and reenlistment bonuses and incentives based on particular MOSs or skills (e.g., flight pay or language proficiency pay). The comparability to AC pay and allowances is important because many other countries have a two-tiered (or more) pay system where the reservists do not receive the same pay or benefits. Even in the United States, one objective of the ongoing duty-status reform process is to reduce the cases where two service members can be performing the same duty but receive different benefits because of the different ways they were called up.

---

[11] For a variety of historical reasons, under the U.S. RC system, inactive-duty training is managed as a standard unit training assembly (UTA) of four hours. This allows time to be bundled into a substantial period but allows the flexibility to move these blocks according to a training schedule. Depending on training requirements, the monthly drill may vary; a unit could add a Friday evening formation and preparation to start a five-period field exercise and then release its personnel at noon on Sunday in a different month for a "UTA-3" weekend.

In addition to the cash compensation for periodic training, RC members earn retirement benefits in the DoD Blended Retirement System (BRS). This includes two types of benefits: a defined-benefit pension that RC retirees may earn after 20 years of qualifying service and a defined-contribution 401(k)-style fund that RC members may contribute to and have the government make certain matching contributions as well.[12] Because the BRS is less than two years old, research predicting its long-term effect on RC recruiting and retention uses models derived from earlier service-member behaviors.[13] Until these models are validated with several more years of behavior under actual BRS options, it will remain difficult to predict how the BRS would interact with the policy innovations discussed in later chapters of this report.

Other Incentives for Reserve Service: Nonmonetary Benefits

In addition to the monetary compensation for service, RC members earn a broad array of nonmonetary benefits and indirect benefits, including but not limited to the following:

- Health care: Active RC members (such as Active Guard/Reserve personnel) are entitled to comprehensive health care from DoD for themselves and their dependents. Selected Reserve members and their dependents are eligible to enroll in TRICARE, a DoD-subsidized health insurance plan. RC members who earn retirement may enroll in TRICARE for Life, a subsidized health insurance plan that turns into a Medicare wraparound plan at age 65.[14]
- Educational benefits: RC members may earn tuition assistance or educational loan repayment in exchange for their service, from either DoD or their state, if they are members of the Army or Air National Guard. In addition to these benefits, RC members who

---

[12] DoD, "Military Compensation," webpage, undated.

[13] See, for example, Beth J. Asch, Michael G. Mattock, and James Hosek, *The Blended Retirement System: Retention Effects and Continuation Pay Cost Estimates for the Armed Services*, Santa Monica, Calif.: RAND Corporation, RR-1887-OSD/USCG, 2017.

[14] TRICARE, "TRICARE Reserve Select," webpage, undated(b).

serve a qualifying active-duty tour may also earn GI Bill benefits through the U.S. Department of Veterans Affairs (VA), up to four years of tuition support, and four years of a housing allowance.

- VA programs: qualifying RC members (generally including those who serve a complete RC term of enlistment or complete an active-duty tour) may be eligible for VA health care services, disability compensation, home loan guarantees, life insurance, or other programs, with eligibility and benefit amounts depending on the RC member's service, health, and other factors.
- Veterans preference for federal employment: the federal government grants a five- or ten-point hiring preference to veterans, including qualifying RC members (generally those with an active-tour or service-connected disability).[15]
- Private-sector employment programs: Since 9/11, a number of private-sector employers have launched veterans-hiring programs targeting separated active and reserve service members on the basis of their military skills, experience, security clearances, and a perception of business value.[16]
- Access to military base facilities: RC members may access military facilities on active bases, such as fitness centers, commissaries, and post exchanges. Purchases made in these facilities are generally free of state sales tax.
- Legal protection from discrimination on the basis of veteran status: Military reservists enjoy certain legal protections that are not afforded to other workers. The Servicemember Civil Relief Act protects reservists against rental property evictions, mortgage foreclosures, insurance cancellations, and government property seizures to pay tax bills during mobilized service.[17] The Uniformed Services Employment and Reemployment Rights Act prohibits

---

[15] Office of Personnel Management (OPM), "Veterans Services," webpage, undated.

[16] Kimberly Curry Hall, Margaret C. Harrell, Barbara Bicksler, Robert Stewart, and Michael P. Fisher, *Veteran Employment: Lessons from the 100,000 Jobs Mission*, Santa Monica, Calif.: RAND Corporation, RR-836-JPMCF, 2014.

[17] See Servicemembers Civil Relief Act, codified at 50 U.S.C. 3901, January 12, 2018, et seq.

employers from discriminating against employees on the basis of military service, including reserve mobilizations, and generally ensures reemployment after periods of activation.[18] State and local statutes provide some additional protections for reservists, as do employer benefits policies that offer differential pay, benefits continuation, or other benefits to reservists during training or mobilization periods. To the extent that discrimination against veterans and reservists remains salient within the workforce,[19] these legal protections may act as a recruiting or retention incentive.

Linking Civilian Skills to Specialties

Implicit in the analysis that follows is the assumption that one's civilian skills—those areas of knowledge, skills, and attributes one acquires through civilian education, employment, and other life experiences—can have some relevance to a military career. For that reason, it may be useful to review how such civilian skills are currently tracked.

At the national level, there is a requirement for all RC members to annually update their Civilian Employment Information. At present, this stand-alone database collects only their name, birth date, DoD identification number, and employer information (job title, employer address, contact information, etc.).[20] Other programs have aimed at gathering more types of data, such as specific skills and certifications, and position types over time.[21]

---

[18] The FY 2006 NDAA also contained narrow provisions for income replacement for certain reservists who experience income loss while involuntarily mobilized. Public Law 109–163, National Defense Authorization Act for Fiscal Year 2006, Sec. 514, 119 Stat. 3136, 3292, January 6, 2006.

[19] Margaret C. Harrell and Nancy Berglass, *Employing America's Veterans: Perspectives from Businesses*, Washington, D.C.: Center for a New American Security, June 2012.

[20] Joint Services Support, "Civilian Employer Information (CEI)," webpage, undated.

[21] For some history of these initiatives and a proposed initiative, as of the early 2000s, see Gregory F. Treverton, David M. Oaks, Lynn Scott, Justin L. Adams, and Stephen Dalzell, *Attracting "Cutting-Edge" Skills Through Reserve Component Participation*, Santa Monica, Calif.: RAND Corporation, MR-1729-OSD, 2003.

The services already use civilian skills to fast-track potential recruits into selected MOSs.[22] These programs differ from the proposals described later in this report in that they generally provide only an alternative path for accession (primarily for the AC) and do not include any other changes to the personnel life cycle that would allow the individual to optimize military service *and* continue in their civilian occupation. However, any improved capability to track the full range of civilian skills within the armed forces could only help the full range of personnel management functions and would directly support each of the constructs proposed in Chapter Seven of this report.

Nonreserve Component Personnel Options Across the Total Force

This report explores alternative approaches for the RCs to access human capital that may be underutilized in current reserve service constructs. Generally speaking, these alternative approaches would be RC solutions—innovative ways to put people in uniform to contribute to military missions. This section provides context for why these approaches merit consideration by describing the policy and practice surrounding the use of other sources of human capital, including DoD civilians and contractors, for functions related to those performed by the RCs. While other elements of the Total Force may benefit from analogous approaches, the scope of this research was intentionally limited to only assess the RCs in this regard.

Policies Governing the Workforce Mix in the Total Force

While in earlier conceptions the Total Force typically referred to the mix of active and reserve forces used to meet military missions, in more recent DoD policy documents (e.g., the *Quadrennial Defense Review 2014*), the Total Force has included military personnel (both AC and

---

[22] For the Army program, see U.S. Army, "Army Civilian Acquired Skills Program ACASP," last modified March 8, 2011.

RC), civilians, and contractors.[23] While the military is often referred to as an entity, the DoDI on the subject makes clear that the RCs are included on the military side of the ledger when it notes,

> If a DoD Component has a military or DoD civilian personnel shortfall, the shortfall is not sufficient justification for contracting an IG [inherently governmental] function. . . . Personnel shortfalls shall be addressed by hiring, recruiting, reassigning military or DoD civilian personnel; authorizing overtime or compensatory time; mobilizing all or part of the Reserve Component (when appropriate); or other similar actions.[24]

Detailed government-wide guidance on the appropriate division of labor between government employees and contractors is provided in "Policy Letter 11-01" from the Office of Federal Procurement Policy in the Office of Management and Budget. This policy letter "provides a single definition of 'inherently governmental function' built around the well-established statutory definition in the Federal Activities Inventory Reform Act (FAIR Act), Public Law 105-270."[25] Appendixes to this policy letter list 24 examples of inherently governmental functions and 9 examples of functions "closely associated" with the performance of inherently governmental functions. It should be noted that, while combat and the command of military forces are listed among the inherently governmental functions included in this document, in general this guidance is intended to delineate between civilian government employees and contractors rather than functions to be divided between service members and contractors.

---

[23] DoD, *Quadrennial Defense Review 2014*, March 4, 2014, p. 47.

[24] DoDI 1100.22, *Policy and Procedures for Determining Workforce Mix*, Washington, D.C., incorporating change 1, December 1, 2017, p. 15.

[25] Office of Management and Budget, "Policy Letter 11-01, Performance of Inherently Governmental and Critical Functions," *Federal Register*, Vol. 76, No. 176, September 12, 2011. A subsequent notice issued in February 2012 clarified that this guidance was intended for both defense and civilian agencies. See Office of Management and Budget, "Policy Letter 11–01, Performance of Inherently Governmental and Critical Functions," *Federal Register*, Vol. 77, No. 29, February 13, 2012, p. 7609.

The principal DoDI that governs the determination of the workforce mix, including what types of roles can be filled by contractors, is DoDI 1100.22, *Policy and Procedures for Determining the Workforce Mix*, issued in 2010.[26] This DoDI defines inherently governmental for the purposes of DoD and codifies statutory language that requires inherently governmental functions to be performed by government employees rather than contractors regardless of whether this is the lowest-cost option.[27] DoDI 1100.22 states,

> In general, a function is IG if it is so intimately related to the public interest as to require performance by Federal Government personnel. IG functions shall include, among other things, activities that require either the exercise of substantial discretion when applying Federal Government authority, or value judgments when making decisions for the Federal Government, including judgments relating to monetary transactions and entitlements.[28]

DoDI 1100.22 details numerous other "manpower mix criteria" that workforce planners are to use when determining the workforce mix. For example,

> even if a function is not [IG] or exempted from private sector performance, it shall be designated for DoD civilian performance . . . unless an approved analysis for either of the following exceptions has been addressed consistent with the DoD Component's regulatory guidelines:
>
> (1) A cost comparison . . . or a public-private competition . . . shows that DoD civilian personnel are not the low-cost provider.

---

[26] DoDI 1100.22, *Policy and Procedures for Determining Workforce Mix*, Washington, D.C., incorporating change 1, December 1, 2017.

[27] Jennifer Lamping Lewis, Edward G. Keating, Leslie Adrienne Payne, Brian J. Gordon, Julia Pollak, Andrew Madler, H. G. Massey, and Gillian S. Oak, *U.S. Department of Defense Experiences with Substituting Government Employees for Military Personnel: Challenges and Opportunities*, Santa Monica, Calif.: RAND Corporation, RR-1282-OSD, 2016.

[28] DoDI 1100.22, 2017, p. 13.

(2) There is a legal, regulatory, or procedural impediment to using DoD civilian personnel. This shall include determinations by Human Resources (HR) officials that DoD civilians cannot be hired in time, or retained to perform the work.[29]

As with Policy Letter 11-01, DoDI 1100.22 is more instructive regarding the distinction between civilian and contractor functions than between military and contractor functions. In addition to a presumption that work is to be performed by government personnel rather than contractors, unless certain conditions are met, there is also a presumption that work should be performed by DoD civilian employees rather than military personnel. DoDI 1100.22 states:

[M]anpower shall be designated as civilian except when one or more of the following conditions apply:

(1) Military-unique knowledge and skills are required for performance of the duties.

(2) Military incumbency is required by law, [executive order], treaty, or international agreement.

(3) Military performance is required for command and control, risk mitigation, or esprit de corps.

(4) Military manpower is needed to provide for overseas and sea-to-shore rotation, career development, or wartime assignments.

(5) Unusual working conditions or costs are not conducive to civilian employment.

Together, these criteria delineate work that is inherently governmental and, essentially, inherently military. Further instruction on the civilian-military workforce mix is included in DoD Directive 1100.4, *Guidance for Manpower Management*, which states,

---

[29] Molly Dunigan, Susan S. Sohler Everingham, Todd Nichols, Michael Schwille, and Susanne Sondergaard, *Expeditionary Civilians: Creating a Viable Practice of Department of Defense Civilian Deployment*, Santa Monica, Calif.: RAND Corporation, RR-975-OSD, 2016.

Manpower shall be designated as civilian except when military incumbency is required for reasons of law, command and control of crisis situations, combat readiness, or esprit de corps; when unusual working conditions are not conducive to civilian employment; or when military-unique knowledge and skills are required for successful performance of the duties.[30]

Despite these guidelines, uncertainty and ambiguity abound, not only regarding the definition of inherently governmental but also regarding the cost-effectiveness of one workforce mix alternative over another. DoDI 7041.04 "establishes policy, assigns responsibilities, and provides procedures to estimate and compare the full costs of active duty military and DoD civilian manpower and contract support," but explicitly excludes Reserve and National Guard manpower from its cost estimates and methodologies.[31] Therefore, it cannot be used here to help understand the trade-offs between our new employment concepts and civilian alternatives. Dunigan et al. (2016) summarize the situation by concluding that "due to the level of interpretation required to assess whether various functions meet these criteria [for the use of contractors], consistent implementation poses a challenge."[32]

One category of contractors that exemplifies the chasm between statutory guidance designed to give priority to government employees and the practice of continuing to rely on contract labor to meet DoD needs are personal services contractors. Per DoDI 1100.22, "A personal services contract is characterized by the employer-employee relationship it creates between the government and the contactor's personnel."[33] On the one hand, the instruction writes that "personal services shall be performed by military or DoD civilian personnel and not contracted unless specifically authorized"; but it goes on to refer-

---

[30] DoD Directive 1100.4, *Guidance for Manpower Management*, Washington, D.C., February 12, 2005.

[31] DoDI 7041.04, *Estimating and Comparing the Full Costs of Civilian and Active Duty Military Manpower and Contract Support*, Washington, D.C., July 3, 2013.

[32] Dunigan, Everingham, et al., 2016.

[33] DoDI 1100.22, 2017.

ence 10 U.S.C. 129b, which provides DoD with broad discretion to enter into personal services contracts including for experts and consultants.[34] The U.S. Government Accountability Office (GAO) has found that the Air Force, Army, and Navy are three of the four agencies across the federal government spending the most on personal services contractors, while also noting inconsistencies in DoD reporting on the use of these contractors.[35] Outside of DoD, authority for circumventing a general prohibition against personal services contracts is found in 5 U.S.C. 3109.[36]

Department of Defense Civilian Employment Alternatives

As shown in Table 2.3, DoD currently directly employs more than 732,000 civilians, largely within the United States. These employees fulfill a broad array of functions across the department and its services, often providing stable personnel alongside military personnel who move frequently as part of the services' normal turnover. DoD plans to expand its civilian workforce to 750,000, largely to meet demands created by the current National Defense Strategy.[37]

Of these, a substantial (but still relatively small) number are in non–full-time forms of employment. As shown in Table 2.4, more than 25,000 DoD employees fell into one of five categories as of March 2018, with full-time seasonal workers accounting for approximately 40 percent of them. These categories are intermittent seasonal, intermittent nonseasonal, part-time seasonal, part-time nonseasonal, and full time.

---

[34] 10 U.S.C. 129b, Authority to Procure Personal Services, January 12, 2018.

[35] GAO, *Federal Contracting: Improvements Needed in How Some Agencies Report Personal Services Contracts*, Washington, D.C., GAO-17-610, July 2017a.

[36] 5 U.S.C. Code 3109, Employment of Experts and Consultants; Temporary or Intermittent, January 5, 1999.

[37] "DoD Projects Growth in Civilian Workforce," *FEDweek*, June 5, 2018.

Table 2.3
U.S. Department of Defense Civilian Employees (Appropriated Fund)

Location	Army	Navy	Air Force	Marine Corps	Other DoD	Total
United States	237,078	187,278	165,554	17,001	98,369	705,280
Abroad	11,586	4,978	3,384	620	6,231	26,799
Total	248,664	192,256	168,938	17,621	104,600	732,079

SOURCE: Defense Manpower Data Center, "Number of Military and DoD Appropriated Fund (APF) Civilian Personnel Permanently Assigned," webpage, June 30, 2018.[1]

[1] Employees paid with nonappropriated funds are excluded from these figures, and include all those paid with revenues from self-funding activities, such as recreation programs like bowling alleys or officer/noncommissioned officer/enlisted clubs.

Department of Defense Contractor Options

DoD contractors fill a wide range of roles, from supplementing DoD's back-office staff to deploying to conflict zones in a support capacity. Types of contracting span from staff augmentation contracting, which typically involves contractors working at DoD facilities while DoD provides most of the other needed inputs to production (e.g., technology and equipment), to complete contracting, which involves a wholesale outsourcing to the contracting firm with DoD only providing high-level oversight and guidance.[38] During the 1990s and early 2000s, DoD aggressively moved to outsource work to the private sector, while the DoD civilian workforce declined by about 38 percent between 1989 and 2002.[39] During this same period, the size of the active military shrunk by about 20 percent.[40] Although most

---

[38] Nancy Young Moore, Molly Dunigan, Frank Camm, Samantha Cherney, Clifford A. Grammich, Judith D. Mele, Evan D. Peet, and Anita Szafran, *A Review of Alternative Methods to Inventory Contracted Services in the Department of Defense*, Santa Monica, Calif.: RAND Corporation, RR-1704-OSD, 2017.

[39] Dunigan, Everingham, et al., 2016.

[40] Under Secretary of Defense for Personnel and Readiness, *Defense Manpower Requirements Report, Fiscal Year 2003*, April 2002.

Table 2.4
Nontraditional Department of Defense Civilian Employment, by Department and Category, March 2018

Department	Intermittent		Part Time		Full Time	Total
	Seasonal	Non-seasonal	Seasonal	Non-seasonal	Seasonal	
Air Force	30	141	713		272	1,156
Army	—	2,554	1,111		765	4,430
Navy	—	228	786	15	1,800	2,829
Other DoD	1,674	2,903	3,497	1,642	7,360	17,076
DoD total	1,704	5,826	6,107	1,657	10,197	25,491

SOURCE: Data from OPM's Fedscope database, as of September 10, 2018.

contractors performed in the United States, providing such services as staff augmentation to headquarters or maintenance support for sophisticated weapon systems, some contractors accompanied U.S. forces into Iraq and Afghanistan to perform services there in support of the military. At the height of the Iraq and Afghanistan wars, contractor personnel (including U.S., host-nation, and third-country nationals) outnumbered U.S. troops, conducting a broad array of functions, including logistics support, translation, construction, and security for diplomatic personnel.[41]

However, a variety of factors led policymakers to seek to swing the pendulum back in the direction of insourcing of defense functions by the late 2000s, including several high-profile incidents in the mid-2000s that raised concerns about contractor accountability and waste, fraud, and abuse, as well as the extent to which contractors were being utilized on the battlefield.[42] These high-profile incidents exemplified

[41] Molly Dunigan, Carrie M. Farmer, Rachel M. Burns, Alison Hawks, and Claude Messan Setodji, *Out of the Shadows: The Health and Well-Being of Private Contractors Working in Conflict Environments*, Santa Monica, Calif.: RAND Corporation, RR-420-RC, 2013.

[42] Moore et al., 2017; Dunigan, Everingham, et al., 2016; Sarah K. Cotton, Ulrich Petersohn, Molly Dunigan, Q. Burkhart, Megan Zander Cotugno, Edward O'Connell, and

broader concerns about overreliance on privatization and outsourcing to the detriment of the organic defense workforce.[43] A 2008 memorandum from the Deputy Secretary of Defense, implementing provisions in the FY 2008 NDAA, stated,

> DoD Components are to ensure consideration is given, on a regular basis, to using DoD civilian employees to perform new functions and functions that are performed by contractors but that could be performed by government employees.[44]

In addition, Congress added reporting requirements for DoD to detail its use of contractors, for example, language in the FY 2008 NDAA that required DoD to publish an *Inventory of Contracted Services*.[45] Since the enactment of this reporting requirement, the methods for generating estimates of obligations for contracted services and the number of contractor full-time equivalents supported by these contracts have evolved, complicating comparisons of contracted services spending over time; however, the general trend has been in the direction of less spending on contracted services (though the drawdown in the wars also complicates the comparison over time since a decline in spending is to be expected in that context).[46]

For our purposes in conceptualizing new forms of RC service, it is important to keep in mind that past practice in DoD may not be the best guide for the future. Especially to the extent that DoD shifted the balance too far in the direction of using contractor labor in the 1990s

---

Michael Webber, *Hired Guns: Views About Armed Contractors in Operation Iraqi Freedom*, Santa Monica, Calif.: RAND Corporation, MG-987-SRF, 2010.

[43] Moore et al., 2017.

[44] Gordon England, Deputy Secretary of Defense, "Implementation of Section 324 of the National Defense Authorization Act for Fiscal Year 2008 (FY 2008 NDAA)—Guidelines and Procedures on In-Sourcing New and Contracted Out Functions," memorandum, April 4, 2008.

[45] Moore et al., 2017.

[46] GAO, *DOD Inventory of Contracted Services: Timely Decisions and Further Actions Needed to Address Long-Standing Issues*, Washington, D.C., GAO-17-17, October 2016.

and early 2000s, including for functions that may have at times crossed the line into inherently governmental, broadening the array of options to leverage human capital within the RC could facilitate a shift toward a more appropriate balance among contractors, civilians, and military personnel.

CHAPTER THREE

Service Demand for Key Types of Personnel

Building on the policy and practice review in the previous chapter, this chapter identifies ways in which current RC and related personnel options seem to be falling short, as revealed through human-resource gaps or challenges at present. Critical to our analysis is the realization that the military manpower system—and, within the scope of this study, specifically the RC—faces challenges in adequately filling specific positions today.[1] If DoD could recruit and retain outstanding individuals for every requirement using the options described above, there would be no need to develop additional options. In practice, each service has some occupational areas for which it is historically hard to recruit and retain personnel, and the services would like to have additional options to provide them. The research team sought to identify current and anticipated manpower shortfalls for each service, specialties that are difficult to recruit and/or retain, and general trends in the military personnel requirements. The intent of this review was not to prepare a comprehensive list or to determine precise requirements. Rather, we sought to gain an overall impression of the types of unmet

---

[1]   In conducting this research, we recognize that historical readiness data are generally not considered to determine whether the prescribed AC/RC force mix and numbers of RC units are realistic. It was outside the scope of this study to gauge the viability of prescribed levels of RC manpower; rather, we were tasked with developing an understanding of how to grow specific areas of expertise within the RC by drawing in those individuals unlikely to serve at all, as well as by enticing existing service members to increase their level of service. Nonetheless, future exploration of this topic may benefit from an in-depth assessment of prescribed RC manpower levels in relation to historical readiness data.

needs the services are facing and help target potential areas that would benefit from alternative work arrangements.

As summarized below and in greater detail in Appendix A, this analysis looks at each service as a whole. Some requirements are best met by AC units and personnel, and we assume that the services will continue developing tools to fill them in this way. However, until such preferred tools are in place, AC requirements will be filled by RC personnel and, as such, should be seen as part of the demand signal for some kind of new RC program.

Early in the research, we identified two distinct ways in which alternative manpower sources could help meet service needs. First, new models could bring into service a number of individuals who are currently unlikely to serve in any capacity. For example, lifting the "Don't Ask, Don't Tell" policy allowed open service to a new segment of the U.S. population. Second, there may be ways in which new models could take currently serving individuals from the minimum level of reserve service, approximately 38 days per year, to a much higher level, such as 180 or more. While the lists of specialties for the two sets of needs overlap (an overall shortage affects both the minimum-duty and extended-duty pools), they will be met by two different populations and require different kinds of changes to policies, regulations, or practices.

The data for different services included to varying degrees quantitative measures of the shortfall. However, we were focused on identifying the *types* of specialties, not measuring the *degree* of shortages or the amount of the gap that a given policy change might close. Matching quantitative degrees of shortfalls to particular policy recommendations that might improve them would be a fruitful avenue for future research.

Workforce Requirements

Peacetime/Support Demand

The RCs follow manpower determination processes prescribed by the services to determine the required size of each functional activity within a unit. This includes the number and type of FTS personnel assigned to a particular unit and the composition of the sustainment "tail" that goes along with the "tooth" of combat forces.

For deployments, DoD requires the military services to present their capabilities in relatively compact, modular packages called unit type codes (UTCs). A UTC may specify equipment, personnel, or both, required to provide a capability. Use of UTCs enables joint combatant commanders to communicate their needs in operational planning documents and actual contingencies in specific, quantifiable, and unambiguous terms. UTCs may be provided by either AC or RC units.

Strategic Demand

In the military RCs, the total number of required units is determined through a comprehensive DoD planning process based on standardized, representative wartime scenarios. The planned active/reserve mix of units is based on cost, baseline readiness, and time-phased availability considerations. Congress responds to DoD inputs in determining the size, funding, and active/reserve mix for the armed forces.

Identifying Shortfalls in Particular Types of Occupational Specialties

Categories of Shortages

To identify specialties suffering from manning shortfalls across the services in line with the study's mandate, the team employed a triangulated approach drawing upon numerous sources of secondary literature and interviews with key service stakeholders. We consulted many pieces of congressional testimony heard by the House and Senate Armed Services Committees, as well as GAO reports on military manpower and shortfalls. This literature assisted in refining our focus on so-called low-density/high-demand occupations. We then conducted interviews with representatives from several of the military services to better refine our understanding of their perceived shortfalls.[2] Appendix A provides

---

[2] Notably, all four services responded to this request for information to some extent; however, the responses from the Air Force and Marine Corps were not sufficiently detailed to

further detail on these data sources and our detailed findings regarding shortage specialties across the services.

We condensed many of the specialties represented in more than one service into larger categories to ease analysis and coordination. These larger categories represent the general characteristics of the associated specialties, although the generalization cannot fully account for the individual nuances of each specialty. In some cases, a category includes other specialties that are not currently hard to fill, but those cases are the exception. The shortage categories we developed were as follows:

- cyber
  - offensive
  - defensive
  - computer/network technicians
- intelligence
- maintenance
- aviation
- medical professionals
- construction
- special operations forces
- investigators
- EOD/CBRN
- linguists
- chaplains
- transportation.

Identifying Common Features in Shortage Specialties

Characteristics of Shortage Specialties

To state the obvious, not all shortages are created equally. For example, the military depends entirely on the private sector to produce the

---

inform our assessments of shortage specialties beyond what we found in the literature and policy review.

nurses, chaplains, and lawyers who are then recruited to fill those military specialties. In other fields, the services train their own personnel but may suffer when those trained individuals leave for better-paying civilian jobs. Appropriate responses to shortages would differ under each condition. To better understand these shortages, we sought information on ten categories of distinguishing characteristics through a literature and policy review, as well as through interviews with representatives from the military services:

1. whether there is a concurrent private-sector shortage of this type of expertise
2. whether there is a concurrent public-sector shortage of this type of expertise
3. the extent of private-sector demand for this expertise
4. whether individuals with such expertise would have been trained in the private sector
5. whether such individuals would have been trained in this skill set in the military
6. whether this specialty entails extensive training requirements
7. whether this specialty entails high risk to the livelihood of the individual performing the work
8. the extent of recruitment standards for this specialty
9. the extent to which this skill is perishable
10. whether this specialty requires extensive use of technology.

Our findings for each shortage specialty are shown in Tables 3.1 and 3.2.

Many of the shortages are not limited to the military, as seen in Table 3.1. For example, it appears that the civilian economy is not producing enough cyber, IT, maintenance, aviation, medical, construction, linguist, and transportation professionals to meet the demands of the private-sector, military, and government agencies. Competition for these professionals is higher as a result. In areas in which the military could produce more talent to meet its needs, there may be other factors that contribute to the shortage. For example, special operators are only produced by the military, but the high risk,

Table 3.1
Demand and Sourcing Characteristics

Specialty Type	Private-Sector Shortage	Public-Sector Shortage	Low/No Private-Sector Demand	Private-Sector Trained	Military Training Pipeline
Offensive cyber		X			X
Defensive cyber	X	X		X	X
Computer/network technicians	X	X		X	X
Intelligence		X			X
Maintenance	X			X	X
Aviation	X	X		X	X
Medical professionals	X	X		X	
Construction	X			X	X
Special operations forces			X		X
EOD/CBRN					X
Linguists	X	X		X	X
Chaplains	X			X	
Transportation	X			X	X

training requirements, and recruitment standards for these positions mean that only a small pool of people will be both eligible and interested. Many of these specialties also either heavily depend on technology or have been or are being transformed by technology. This trend affects specialties in different ways, but it may create new opportunities to be creative in mitigating shortfalls. While we did not study the various impacts of specific shortfalls, we noted the characteristics of particular shortage specialties—particularly regarding time spent on the job and location of the position—to help identify workarounds in potential new workforce constructs that might be utilized to increase available manpower for the RC in such specialty areas.

Table 3.2
Specialties Characteristics

Specialty Type	High Training Requirements	High Risk	High Recruitment Standards	Highly Perishable Skill	Extensive Use of Technology
Offensive cyber	X			x[a]	X
Defensive cyber	X			x[a]	X
Computer/ network technicians					X
Intelligence					X
Maintenance	X				X
Aviation		X		X	X
Medical professionals					X
Construction		X			
Special operations forces	X	X	X		x[a]
EOD/CBRN		X			x[a]
Linguists	X			x[a]	
Chaplains					
Transportation		X			x[a]

[a] x indicates a particularly variable condition. For example, language skills acquired by adults in a school setting are generally very perishable, but the skills of heritage speakers require little sustainment training. For some cyber skills, proficiency and active engagement on a specific network improves performance.

CHAPTER FOUR

Identifying Segments of the Labor Force Underrepresented in Military Service

With so many career fields showing systemic patterns of shortfalls within the services, it seems unlikely that the solution is simply to improve recruiting incrementally. A radical solution calls for an examination of new areas from which to recruit or an exploration of ways to increase the level of participation among those who have already chosen to serve, with a particular focus on how an expanded RC population could more adequately supplement the AC. This chapter examines policy and regulatory limits that may be relevant to mitigating recruitment shortfalls and develops a conceptual framework to guide further examination of other key limiting factors related to participation in the RC. The conceptual framework integrates population-level data on civilian occupations to highlight groups that may face barriers or other challenges to RC participation, as well as potentially untapped groups with desired skill sets within the larger U.S. population.

Policy and Regulatory Limits to Service

To participate in the RCs, individuals must first meet basic requirements in several broad categories. First, per DoD regulations, there are detailed personal history and physical requirements that must be met (with some variation by branch of service). Second, after completing initial training (8 to 13 weeks) and technical training (the duration of which varies widely based on specialty), traditional reservists are required at minimum to participate in reserve activities 1 weekend per month plus 2 weeks per year at a base that is a "reasonable commute" from where they live.[1] Taken broadly, the participation requirements may be characterized as privileging certain personal attributes, or characteristics of their time- or location constraints, over others. Therefore, these sets of requirements as currently defined may exclude certain types of individuals who otherwise have valuable, in-demand skills to contribute to RC service. In this section, we examine ways in which these requirements may limit or preclude certain types of individuals or occupations from participation in the RC, identify specific groups for which these requirements may limit participation, and provide estimates of the sizes of some of these key populations. We do not argue that any or all of these restrictions *should* be changed, but we do note their existence and their effects on the recruitable population.

Personal Characteristic Requirements for Participation in the U.S. Military

While there are slight variations between each of the military services, applicants cannot have any serious law violations or drug use; no history of serious health problems; meet age, height, and weight standards; score sufficiently on an aptitude test (Armed Services Vocational Aptitude Battery); and pass a physical exam. While the standards may vary and change over time, the criteria apply in some way to all military service, not only that in an RC.

Height-for-weight and body-fat requirements are increasingly limiting eligibility; the high and rising prevalence of overweightness and obesity in the civilian population reduces the available pool of recruits.[2] In 2007–2008, 11.7 percent of military-age men and 34.7

---

[1]   In practice, many RC members, particularly those in the more senior noncommissioned officer and officer grades, are voluntarily assigned to positions beyond that commuting distance, most often to accept a promotion or an opportunity to serve in a key developmental position.

[2]   Katherine M. Flegal, Deanna Kruszon-Moran, Margaret D. Carroll, Cheryl D. Fryar, and Cynthia L. Ogden, "Trends in Obesity Among Adults in the United States, 2005 to

percent of women exceeded the standards.[3] Furthermore, 23 percent of all medical disqualifications in 2006 were due to height and weight.

Beyond physical condition requirements, some health condition requirements may also increasingly limit participation. One example of such a disqualifying health condition is that individuals cannot have used sleep aids or have had difficulty sleeping (chronic insomnia) more than three times per week during the past three months. Data from 2005–2010 indicated that 4 percent of Americans age 20 and older had used prescription sleep aids in the previous month,[4] that there was a 293-percent increase in the number of sleep medication prescriptions from 1999 to 2010,[5] and that 20–25 percent of men and 15–18 percent of women suffered from sleep disturbances more than six times in a two-week period.[6] Workers in occupations that require night, rotating, or extended shifts, such as police, firefighters, and other first responders, have an increased prevalence of sleep disorders.[7] Note that condi-

---

2014," *Journal of the American Medical Association,* Vol. 315, No. 21, 2016; Doosup Shin, Chandrashekar Bohra, Kullatham Kongpakpaisarn, and Eun Sun Lee, "Increasing Trend in the Prevalence of Abdominal Obesity in the United States During 2001–2016," *Journal of the American College of Cardiology,* Vol. 71, No. 11, supplement, March 2018.

[3]   J. Cawley and Catherine Maclean, "Unfit for Service: The Implications of Rising Obesity for US Military Recruitment," *Health Economics,* Vol. 21, No. 11, November 2012.

[4]   Yinong Chong, Cheryl D. Fryar, and Qiuping Gu, "Prescription Sleep Aid Use Among Adults: United States, 2005–2010," National Center for Health Statistics Data Brief No. 127, August 2013.

[5]   Earl S. Ford, Anne G. Wheaton, Timothy J. Cunningham, Wayne H. Giles, Daniel P. Chapman, and Janet B. Croft, "Trends in Outpatient Visits for Insomnia, Sleep Apnea, and Prescriptions for Sleep Medications Among US Adults: Findings from the National Ambulatory Medical Care Survey 1999–2010," *Sleep,* Vol. 37, No. 8, 2014.

[6]   Michael A. Grandner, Jennifer L. Martin, Nirav P. Patel, Nicholas J. Jackson, Philip R. Gehrman, Grace Pien, Michael L. Perlis, Dawei Xie, Daohang Sha, Terri Weaver, and Nalaka S. Gooneratne, "Age and Sleep Disturbances Among American Men and Women: Data from the U.S. Behavioral Risk Factor Surveillance System," *Sleep,* Vol. 35, No. 3, March 1, 2012.

[7]   Al Pack and G. W. Pien, "Update on Sleep and Its Disorders," *Annual Review of Medicine,* Vol. 62, 2011; Shantha Rajaratnam, Laura K. Barger, Steven W. Lockley, Steven A. Shea, Wei Wang, Christopher P. Landrigan, Conor S. O'Brien, Salim Qadri, Jason P. Sullivan, Brian E. Cade, Lawrence J. Epstein, David P. White, and Charles A. Czeisler, "Sleep Dis-

tions that may prevent individuals from enlisting do not necessarily lead to removal from service or prevent personnel from deploying. For example, although there has been a marked increase in rates of sleep apnea and insomnia among active-duty military personnel since 2000, these rates are still lower than among the U.S. civilian population.[8]

Additional restrictions include personal features, such as the presence of tattoos. Personnel are not permitted to have tattoos on the head, face, neck, wrists, hands, or fingers, and they also may not have more than four tattoos below the knee or between the elbow and wrist.[9] While statistics on tattoos among the general U.S. population are relatively scarce, their prevalence appears to vary across occupations, with agriculture (22 percent) and hospitality, tourism, and recreation (20 percent) reporting the highest rates.[10] Hand and arm tattoos on professional chefs appear to be particularly common.[11] Overall, 38 percent of Americans age 18–29 report having tattoos, and 18 percent of those have six or more, although only 30 percent reported that their tattoos are usually visible.[12]

---

orders, Health and Safety in Police Officers," *Journal of the American Medical Association*, Vol. 306, No. 23, 2011; Laura K. Barger, Shantha M. W. Rajaratnam, Wei Wang, Conor S. O'Brien, Jason P. Sullivan, Salim Qadri, Steven W. Lockley, and Charles Czeisler, "Common Sleep Disorders Increase Risk of Motor Vehicle Crashes and Adverse Health Outcomes in Firefighters," *Journal of Clinical Sleep Medicine*, Vol. 11, No. 3, March 2015.

[8] Vincent Mysliwiec, Leigh McGraw, Roslyn Pierce, Patrick Smith, Brandon Trapp, and Bernard J. Roth, "Sleep Disorders and Associated Medical Comorbidities in Active Duty Military Personnel," *Sleep*, Vol. 36, No. 2, 2013.

[9] Army Regulation 670–1, *Uniform and Insignia Wear and Appearance of Army Uniforms and Insignia*, Washington, D.C., March 31, 2014.

[10] Wendy Heywood, Kent Patrick, Anthony M. A. Smith, Judy M. Simpson, Marian K. Pitts, Juliet Richters, and Julia M. Shelley, "Who Gets Tattoos? Demographic and Behavioral Correlates of Ever Being Tattooed in a Representative Sample of Men and Women," *Annals of Epidemiology*, Vol. 22, No. 1, January 2012; Statista, "Do You Have a Tattoo? (by Occupation)," 2013 data, webpage, undated.

[11] See, for example, Jessica Gelt, "The Latest Trend Among Chefs: Food Tattoos," *Los Angeles Times*, April 27, 2012; and Gretchen McKay, "Illustrated Chefs: Why Kitchen Artists Are Big on Tattoos," *Pittsburgh Post-Gazette*, June 19, 2011.

[12] Paul Taylor and Scott Keeter, eds., *Millennials: A Portrait of Generation Next*, Washington, D.C.: Pew Research Center, February 2010.

Another personal characteristic that potentially precludes service relates to Army, Air Force, and Marine Corps grooming regulations prohibiting facial hair.[13] Some civilians, such as members of the Hasidic Jewish community, are required to have beards for religious reasons. Although this specific population is relatively small and represents a minority within the small U.S. Jewish population (the total Hasidic population of the United States was estimated to be 180,000 in 2000), this restriction could contribute to both overall low recruiting of this population and the Jewish chaplain shortage in the military.[14] That said, as of 2014 DoD has begun to consider religious exemptions for certain requirements, such as hair and jewelry.[15]

There are also enlistment restrictions related to dependents.[16] Single parents (unmarried, divorced, or widowed) cannot enlist unless they relinquish custody of their children or get an approved family care determination. This restriction alone represents an additional hurdle for roughly 20 million Americans. Transferring custody of children requires potentially burdensome legal proceedings to obtain a court order, possibly requiring a lawsuit and provision of child support. Furthermore, requirements vary by service. For example, the Navy requires a waiver for any applicant with more than one dependent, including a spouse.[17] The Marine Corps requires a waiver for applicants with any dependent under 18, and the Air Force conducts a financial eligibility determination for all applicants with dependents.[18] The Army requires

---

[13] Army Study Guide, "Hair Standards," webpage, undated.

[14] Joshua Comenetz, "Census-Based Estimation of the Hasidic Jewish Population," *Contemporary Jewry*, Vol. 26, 2006; Union for Reform Judaism, "Support for Jewish Military Chaplains and Jewish Military Personnel and Their Families," resolution, 2005.

[15] DoDI 1300.17, *Accommodation of Religious Practices Within the Military Services*, Washington, D.C., incorporating change 1, January 22, 2014.

[16] DoDI 1304.26, *Qualification Standards for Enlistment, Appointment, and Induction*, Washington, D.C., incorporating change 2, April 11, 2017.

[17] Navy CyberSpace, "Navy Dependency Waiver," webpage, June 17, 2018.

[18] Marine Corps Order P1100.72C, *Military Personnel Procurement Manual, Volume 2: Enlisted Procurement*, June 18, 2004; Air Force Instruction 36-2906, *Personal Financial Responsibility*, July 30, 2018.

a waiver when an applicant has two or more dependents in addition to a spouse.[19]

Despite the substantial list of characteristics that can preclude individuals from service, it is possible to obtain waivers for many of them. Indeed, in 2007, about one in five new military recruits had some kind of eligibility waiver.[20] This suggests that, in general, personal characteristics may be less systematically restrictive than other factors, such as time constraints.

Time and Location Constraints

Given that many personal characteristic requirements that may otherwise preclude RC participation may be waived, we now examine the potential time- and location-based constraints that could make RC participation difficult. Participation one weekend per month and two weeks per year requires service members to be physically present at their unit training site during these times. For some individuals, personal commitments, work-related travel, living in rural areas, or other factors may make physical presence a challenge. We conceptualize time constraints largely with respect to characteristics of work schedules and need for work-life balance.

Military personnel must also be available for deployment. This should always be an assumption when it comes to military service, but this policy is accompanied by a mandate for corrective action. A February 14, 2018, memo issued by the Under Secretary of Defense for Personnel and Readiness states, "Service members who have been non-deployable for more than 12 consecutive months, for any reason, will be processed for administrative separation." (There is a blanket excep-

---

[19] U.S. Army Recruiting Regulation 6601-566, *Personnel Procurement: Waiver, Future Soldier Program Separation, and Void Enlistment Processing Procedures*, May 31, 2006; Army Regulation 601-210, *Regular Army and Reserve Components Enlistment Program*, Washington, D.C., August 31, 2016; Army Regulation 135-18, *The Active Guard Reserve Program*, Washington, D.C., September 29, 2017.

[20] Phillip Carter, Katherine Kidder, Amy Schafer, and Andrew Swick, *AVF 4.0: The Future of the All-Volunteer Force*, working paper, Washington, D.C.: Center for a New American Security.

tion for pregnant and postpartum members, and there are procedures for *temporary* waivers.)[21]

One initial hypothesis of our research was that policies and regulations originally intended to give structure to RC participation might, in fact, amount to barriers that make participation by certain demographic or occupational groups less likely. We used three nationally representative data sources to help identify occupational categories with substantial potential barriers to service: the American Working Conditions Survey (AWCS), the General Social Survey (GSS), and the American Community Survey (ACS). As described in further detail below, the descriptive results from the GSS and ACS analyses taken together suggest that factors associated with work schedules, including irregular/on-call hours, split/rotating shifts, and frequent extra hours may be hard to reconcile with RC service. Similarly, occupations with particularly taxing mental or physical requirements appear to be less conducive to RC service.

However, there are exceptions to these general findings, cases of occupations with scheduling and work-life balance characteristics that would seem to make RC less likely but that in fact have higher-than-average RC participation rates. This suggests that other factors, such as being in an occupation involving skills that translate well to military service or having a higher motivation to serve, can outweigh factors that make it more logistically difficult. Because we did not have the resources to conduct our own survey, we are unable to untangle these factors, to identify more precisely the extent to which the hypothesized scheduling or geographic obstacles to RC service are indeed reducing participation, or to directly test whether one or more of the pilot proposals we describe later in this report would be effective in overcoming these obstacles.

Figure 4.1 shows the four time-location obstacle combinations that individuals may experience, according to our literature review and interviews as well as patterns in the AWCS, GSS, and ACS data sets (as elaborated further below). Green shading (cell A) indicates minimal

---

[21] Wilke, 2018.

or no constraints to RC participation. Yellow (cells B and C) indicates that there are potential constraints in one of the dimensions, and red (cell D) indicates constraints along both dimensions. Occupations in the yellow and red cells are those for which we attempted to identify alternative service models that could mitigate the time or location obstacles defining that group. Note that the same constraints may affect an employee in any occupation requiring extensive travel, weekend work, or shift work, as well as those with parenting and caregiving responsibilities. They are also not necessarily mutually exclusive; for example, a combination of work-related and parenting/caregiving demands that could be barriers to RC service is common.

Cell A includes individuals for whom neither time nor location are likely obstacles to participation in the RC; current RC requirements do not present barriers to participation on these dimensions. Cell B includes individuals for whom committing the time to participate in RC activities likely is not problematic but who may face challenges in accessing the fixed training base location. Individuals in this group include those whose jobs require them to be geographically mobile or make them otherwise unlikely to be in their usual home location at the required training times. For example, truck drivers may be away from home for extended periods, and overseas language teachers may spend months at a time in a location that makes regular participation in geographically bound training impossible.

Cell C, in which time but not location is problematic, consists of a varied population that may face difficulty committing to a minimum monthly weekend or annual two-week block of service. Multiple aspects of civilian occupation work schedules could contribute to this conflict. Some occupations require weekend, on-call, or shift work, and seasonal or block employment may not permit workers to take the necessary time off to participate in the RC, such as for agricultural work during planting and harvesting season, athletes and sports industry workers during game season, or tax professionals and accountants during tax season. Related to this, occupations with frequent tight deadlines may demand that individuals regularly work outside of regular hours or days, such as lawyers or those in IT and computer programming fields. Another group with potential time constraints

Figure 4.1
Conceptual Chart of Time-Location Obstacles to Reserve Participation

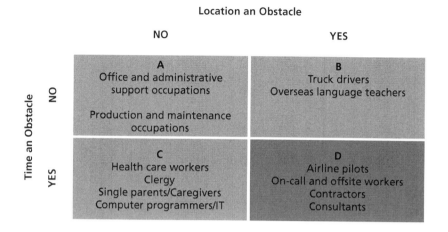

to reserve participation are those who are self-employed, regardless of occupation; the demands of running a small business may preclude participation in the reserves. Similarly, individuals with family care responsibilities, for children or the elderly, for example, may be unable to make time for RC participation. Others, such as health care workers or firefighters, may have both routine weekend shifts but also on-call responsibilities.

Cell D consists of people who face both time and location obstacles and who may face the greatest challenges to participation in the reserves. These individuals face many of the same time or work-life balance constraints as in Cell C, further complicated by potentially limited geographic access to a home training base. For example, some occupations that may belong to this group are pilots and flight crews, contractors or consultants with limited control over when *and* where they work, oil rig workers, and deep-sea fishers.

To be clear, we are not discussing formal or explicit barriers placed by employers on reserve service. The Uniformed Services Employment and Reemployment Rights Act of 1994 provides employment protections and remedies to veterans and service members (including reservists). The act's provisions stipulate that employers must permit reservists to take time off from work to participate in military activities without any penalty or retaliation, provided employees provide proper notice to their employers. In theory, this implies that no degree of work schedule inflexibility should preclude participation in the reserves. In practice, some friction remains, and neither employees nor employers can be expected to continue a relationship in which military service demands accommodations on a continuous or routine basis.[22] In this project, we will focus on occupations where such conflicts are inherent to the civilian occupation or employment arrangement and therefore beyond the *practical* application of the Uniformed Services Employment and Reemployment Rights Act.

The American Working Conditions Survey Analysis

Next, we explore these conceptual obstacles using several nationally representative data sources—the AWCS, GSS, and ACS—to help identify occupational categories that report substantial potential for barriers to service. Using the ACS data, we estimated the number of individuals currently employed in selected occupations with particularly relevant skills but substantial potential barriers.

First, we looked at the AWCS data to obtain an initial assessment of links between work schedules, work-life balance, and occupational groups. The AWCS was fielded to RAND American Life Panel (ALP) respondents in 2015.[23] The AWCS collected information on several dimensions of working conditions and work-life balance, among other topics. The sample of 3,131 respondents age 18–71 included 2,071 currently employed workers. After weighting, the AWCS compares well with the Current Population Survey on the proportion of Americans

---

[22] Harrell and Berglass, 2012.

[23] The ALP is a nationally representative sample of U.S. residents who have agreed to participate in regular online surveys. Respondents who do not have a computer at home are provided both a computer and internet access to ensure that the panel is representative. Since its inception in 2006, the ALP has fielded more than 500 surveys on a wide variety of topics, including health, employment, and retirement.

working for pay (conditional on working for pay), the proportions self-employed and working part time (defined as less than 35 hours per week), and average hours worked per week. In the AWCS, three-quarters of men and just over two-thirds of women age 18–71 worked for pay in the third quarter of 2015.[24]

Variables of Interest

Here, we examined a series of questions related to aspects of work schedules and work-life balance among the currently employed, grouping across occupations, to explore whether there are types of occupations that are associated with characteristics that may impede participation in reserve service. Specifically, we investigated whether respondents reported having low levels of control over their work schedules and whether they reported higher levels of work-life conflict, both of which could make reserve participation more difficult according to the conceptual framework in Figure 4.1.

1. How many days per week do you usually work in your main paid job?
2. Do you work the same number of days every week?
3. How are your working time arrangements set?
   - They are set by the company/organization with no possibility for changes.
4. In general, how would you describe your hours on your main job?
   - Predictable seasonal work during the year.
   - Unpredictable or irregular work (e.g., unpredictable periods without work, layoffs, and/or sporadic hours).
5. Does your job involve work to tight deadlines?
6. In general, do your working hours fit in with your family or social commitments outside work?
7. Thinking about both your commitments at work and outside of work, please select the response that best describes your situa-

---

[24] Nicole Maestas et al., *The American Working Conditions Survey Data: Codebook and Data Description*, Santa Monica, Calif.: RAND, TL-269-APSF, 2017.

tions. How often, in the last three months, has it happened that you were too tired for activities in private life?

Results

The AWCS sample size requires that occupations be collapsed to provide cells of sufficient sizes for analysis. We compare weighted responses to the working conditions survey questions across the following occupational groups: administrative support and retail sales; executive, administrative, and managerial; farm operators and managers; financial sales and related; food preparation and service; management related; professional specialty; technicians and related support; transportation, construction, mechanics, mining, and agriculture. Complete results are presented in Appendix B, Table B.1. There are several notable differences in reported proportions across some of the occupational groups:

- First, the "food preparation and service" workers report substantially higher incidence (45 percent, which is four times the average across all occupations) of usually working six or seven days per week (Question 1), which could make participation in the reserves one weekend per month difficult.[25]
- Second, nearly half of the "administrative and retail" workers report having tight deadlines (Question 5) "all" or "almost all of the time," and nearly a third report being too tired for activities in private life "always" or "most of the time" (Question 7). Each of these responses is roughly 50 percent higher than the average across all occupations.

---

[25] Complementary data from the 2016 American Time Use Survey, published by the Bureau of Labor Statistics (BLS), identifies "sales and related" and "service" occupations as substantially more likely than others to involve working on an average weekend or holiday (47.5 percent and 40.2 percent, respectively). Similarly, self-employed workers were more likely than wage and salary workers to regularly work on weekends (37.8 percent). BLS, "Employed Persons Working on Main Job and Time Spent Working on Days Worked by Class of Worker, Occupation, Earnings, and Day of Week, 2017 Annual Averages," press release, June 28, 2018f.

- Third, both the "farm operators and managers" and "transportation, construction, mechanics, mining, and agriculture" workers report high prevalence of irregular days in their typical work schedules (Question 2) and seasonal work (Question 4—roughly twice the average across all occupations for both items). Further, they both report the highest scores on conflict between working hours and family/outside commitments (Question 6) and in being too tired for activities outside of work (Question 7) among all occupations.

Of the questions related to working conditions in the AWCS, only one item did not identify specific occupations with relatively high mean differences—if working time arrangements are set by the company with no possibility for changes (Question 3). The conceptual model in Figure 4.1 suggests that scheduling inflexibility could limit ability to participate in the RC; while inflexible schedules were not substantially more common in one occupation compared with the overall average, working six or seven days a week (precluding weekend availability), differing numbers of days every week, seasonal work, and frequent tight deadlines all distinguished various occupations. The two aspects of work-life balance (Questions 6 and 7) further identified several occupations as substantially less balanced than the average. Together, these identified occupations may face challenges to RC participation consistent with Figure 4.1. While these differences are largely suggestive, we can turn to a larger data set to expand the picture.

General Social Survey Analysis

As part of our investigation into groups of people who may be underutilized in traditional reserve service constructs because of scheduling or other work-life conflicts, we analyzed data from the GSS Quality of Working Life (QWL) module.[26] The GSS is a nationally representative

---

[26] National Opinion Research Center, "General Social Surveys—Quality of Working Life Module, 1972–2014: *Cumulative Codebook*," Chicago, Ill.: University of Chicago, June 2017.

sociological survey fielded every two years by the National Opinion Research Center (NORC) at the University of Chicago, with funding from the National Science Foundation (NSF). Since 2002, NSF has joined with the National Institute for Occupational Safety and Health to support a quadrennial QWL module, which includes a variety of questions that shed light on the share of the population with various types of work-life conflicts that we believe could affect an individual's ability to participate in the reserves.

While some questions relating to the quality of working life were asked in other years the GSS was administered, the full set of questions was only asked in 2002, 2006, 2010, and 2014. Only individuals working full time, part time, or temporarily not at work were eligible to answer the QWL questions. In our analysis, we pool across the years the QWL survey was fielded. While this may result in some loss of granularity in terms of changes in the quality of working life over time, pooling improves sample size and facilitates a look at groups of occupations that may experience particular work-life challenges. The analyses are weighted.

Variables of Interest

We reviewed the GSS QWL module codebook to identify variables that plausibly could be used to estimate the share of workers with work-life obstacles to reserve service. We focused on questions that impinged on scheduling and the amount of useful time an individual has outside of work, while bypassing other questions (e.g., those related to job satisfaction and job security). We selected eight QWL questions and coded them into nine variables as follows:

1. Which of the following best describes your usual work schedule?
   - *Variable 1: Irregular shift/on-call schedule*
   - *Variable 2: Irregular shift/on-call or split shifts or rotating shifts schedule*
2. How many days per month do you work extra hours beyond your usual schedule?
   - *Variable 3: Works extra hours ten or more days per month*

3.   When you work extra hours on your main job, is it mandatory (required by your employer)?
   • *Variable 4: Extra hours are required by employer*
4.   How hard is it to take time off to take care of personal or family matters?
   • *Variable 5: Somewhat hard or very hard to take time off to take care of personal or family matters*
5.   How often do the demands of your job interfere with your family life?
   • *Variable 6: Demands of job often interfere with family life*
6.   How often have you felt used up at the end of the day?
   • *Variable 7: Often or very often felt used up at the end of the day*
7.   How often have you come home from work too tired to do the chores that need to be done?[27]
   • *Variable 8: Comes home several times a week too tired to do the chores*
8.   How often has it been difficult for you to fulfill your family responsibilities because of the amount of time you spend on your job?
   • *Variable 9: Difficult to fulfill family responsibilities several times a week because of amount of time spent on job*

Results

We analyze the GSS data both in the aggregate for all workers who answered the QWL module questions and broken down by categories of occupations.[28] The aggregate shares (across all occupations) provide insight into the shares of the overall civilian workforce that experience

---

[27] Questions 7 and 8, corresponding to Variables 8 and 9, were asked in the 2002 survey only.

[28] The complete list of census occupational categories identified by BLS is as follows: architecture and engineering; arts, design, entertainment, sports, and media; building and grounds cleaning and maintenance; business and financial operations; community and social service; computer and mathematical; construction and extraction; education, training, and library; farming, fishing, and forestry; food preparation and serving related; health care practitioners and technical; health care support; installation, maintenance, and repair; legal; life, physical, and social science; management; office and administrative support; per-

these types of work-life balance challenges, which we hypothesize may make them less likely to participate in the reserves. These characteristics potentially situate individuals and occupational situations where time is an obstacle to RC participation but location is not an obstacle (Cell C of the conceptual model of obstacles to RC participation, Figure 4.1). Given the time requirements for participation in the RC, it is possible that occupations with greater time demands, less predictable schedules, less control over scheduling, or lower levels of work-life balance will be less able to participate. The occupational breakdowns may be useful in identifying types of jobs that are particularly associated with work-life conflict. To the extent that jobs with high rates of work-life conflict involve skills that the reserves are seeking to attract, designing innovative ways to access human capital could improve rates of reserve participation for workers who hold these jobs. We again caution that we are unable to identify the extent to which these hypothesized barriers to RC service are in fact reducing participation, though pairing the results from the GSS analysis with the ACS analysis (below) provides suggestive evidence of a possible connection between these scheduling and work-life balance obstacles and reduced rates of RC participation for some occupations.

We present the distributions of each of the GSS characteristics by occupation, as well as the mean across all occupations, in Table 4.1. Our key descriptive findings from the GSS data, abridged to focus on the specific occupational groups in Table 4.1, are as follows:[29]

- Three occupational groups have shares above the overall averages for seven of nine variables: health care practitioners and technical occupations, protective service occupations, and transportation and material moving occupations. Together, these three occupational groups contained an estimated 22.0 million workers in 2016, according to the ACS.

---

sonal care and service; production; protective service; sales and related; transportation and material moving. See BLS, "Census 2010 Occupation Codes," undated(a).

[29] The complete table is presented in Table B.2 in Appendix B.

- Another two groups have shares above the average for six of nine variables: community and social service occupations and construction and extraction occupations. These two occupational groups contained an estimated 10.3 million workers in 2016.
- Rounding out the occupational groups with at least five variables with higher-than-average shares are education, training, and library occupations, food preparation and serving-related occupations, and management occupations—together containing a further 33.8 million.
- Combined, these eight occupational groups contained 43 percent of all civilian employed workers in 2016.
- Further, in some cases, an occupational group did not have above-average shares for a majority of all questions but received the highest shares for individual questions. For example, the occupational group with the highest reported share with irregular/on-call schedules is the arts, design, entertainment, sports, and media occupations group (21 percent, 3.1 million workers in 2016). This occupational group also has among the highest shares when including split and rotating schedules in addition to irregular/on-call schedules (30 percent).

A more detailed look at which individual occupations within the occupational groups are driving these findings is limited by the small sample sizes for individual occupations in the GSS QWL data, even when pooling across the years. As a result, these findings should be interpreted with caution and in tandem with other quantitative and qualitative evidence regarding which types of jobs may pose scheduling obstacles to RC participation. Nonetheless, the list of the individual occupations with the highest reported shares of workers with irregular or on-call schedules, for example, is intriguing in its breadth and variety across types of occupations. Occupations with the highest share of workers reporting irregular or on-call schedules include the following:

- sailors and marine oilers
- flight attendants
- aircraft pilots and flight engineers

- miscellaneous health care support occupations
- tax preparers
- clergy
- real estate brokers and sales agents
- chefs and head cooks
- hazardous materials removal workers
- cabinetmakers and bench carpenters
- food service managers
- architects
- construction and building inspectors
- physicians and surgeons
- medical scientists.

When including split and rotating schedules, the following additional individual occupations also have among the highest share of workers with irregular or on-call schedules:

- firefighters
- geological and petroleum technicians
- agricultural inspectors
- first-line supervisors of firefighting and prevention workers
- emergency medical technicians and paramedics
- first-line supervisors of police and detectives.

American Community Survey Analyses

The GSS and AWCS identify groups of occupations with work-life characteristics that could be problematic for reserve service, according to our conceptual model. Given this, we can next examine whether these occupations also report substantially lower levels of RC participation, as we might expect, all else being equal, which would be consistent with our model. Having identified occupations with characteristics that are conceptually linked to barriers to RC participation, do we also see lower RC participation for these same occupations, as predicted by our model? As noted previously, this does not consider other factors,

Table 4.1
Work-Life Characteristics from the General Social Survey Quality of Working Life

Occupational Group	Irregular Shift/ On-Call Schedule	Irregular Shift/ On-Call or Split Shifts or Rotating Shifts	Works Extra Hours 10+ Days per Month	Extra Hours Required by Employer	Somewhat Hard or Very Hard to Take Time Off	Job Demands Often Interfere with Family Life	Often Feels "Used Up" at the End of the Day	Comes Home Too Tired for Chores	Difficult to Fulfill Family Responsibilities
All occupations	9%	17%	23%	27%	27%	13%	41%	28%	11%
Community and social service	14%	20%	26%	24%	23%	17%	44%	33%	0%
Computer and mathematical	6%	9%	29%	22%	14%	8%	34%	56%	31%
Construction and extraction	10%	12%	27%	34%	25%	13%	43%	19%	16%
Education, training, and library	4%	6%	37%	18%	38%	11%	45%	34%	19%
Food preparation and serving related	10%	32%	12%	25%	33%	11%	48%	28%	19%
Health care and technical	9%	20%	20%	27%	41%	14%	45%	43%	23%
Management	10%	15%	40%	26%	24%	17%	40%	29%	11%
Protective service	10%	40%	14%	44%	33%	26%	41%	34%	16%

Table 4.1—Continued

Occupational Group	Irregular Shift/ On-Call Schedule	Irregular Shift/ On-Call or Split Shifts	Works Extra Hours 10+ Days or Rotating Shifts	Extra Hours per Month	Extra Hours Required by Employer	Somewhat Hard or Very Hard to Take Time Off	Job Demands Often Interfere with Family Life	Often Feels "Used Up" at the End of the Day	Comes Home Too Tired for Chores	Difficult to Fulfill Family Responsibilities
Transportation and material moving	16%	26%	22%	37%	39%	15%	44%	23%	14%	

SOURCE: NORC, University of Chicago, General Social Survey data (abridged).

such as being in an occupation involving skills that translate well to military service or having a higher propensity to serve. However, consister̵cy between the observed occupational characteristics and actual RC participation by occupation helps validate our conceptual model.

The ACS is a large, ongoing survey conducted by the U.S. Census Bureau that sampled respondents are required by law to complete, which should maximize participation rates and minimize undercoverage.[30] The ACS collects data each year to bridge intercensal periods and provide detailed information about the population. The ACS includes detailed occupation codes, as well as participation in military service. We considered respondents who reported ever being on "active duty for training in Reserves/National Guard" to have been in the RC; in the unlikely case they had been activated for duty but never purely for training, we were unable to identify their reserve service. As a result of these cases, the ACS may underestimate the reserve population. Furthermore, we were unable to identify the period of reserve service, which hampered our interpretation somewhat. However, as shown in Table 4.2, we produced a series of estimates to account for this limitation. Despite these limitations, the ACS offers us an unmatched opportunity to compare occupations and reserve service. Note that the ACS estimates are period estimates; when precision of estimates is more important than currency of estimates, the U.S. Census Bureau recommends using the five-year ACS estimates rather than the one-year estimates. Thus, we relied on the five-year ACS estimates (2011–2015, N = 15.6 million) of reserve participation by occupational group.[31] Further, while the ACS includes specific occupation, we have collapsed occupations into their larger occupational groups for parsimony. However, we do discuss several specific occupations in a later section.

Table 4.2 presents the weighted percentage of currently employed workers in selected occupational groups that reports reserve service. A table presenting expanded information for all occupational groups can

[30] The relevant laws are 18 U.S.C. 3571 and 3559, which amends 13 U.S.C. 221.

[31] Michael Beaghen and Lynn Weidman, *Statistical Issues of Interpretation of the American Community Survey's One-, Three-, and Five-Year Period Estimates*, Washington, D.C.: U.S. Census Bureau, 2008.

be found in Table B.3 in Appendix B. To account for potential gender differences across occupations, we present separate estimates for men and women. The first two columns are based on workers of all ages. Next, we focus on workers age 40 and younger, to capture individuals more likely to participate in the reserves rather than including individuals who may have entered a career only subsequent to service in the reserves.

The table indicates that roughly 1.5 percent and 0.5 percent of all currently employed men and women have participated in the RC, respectively, an average of 1.0 percent overall. When we focus on those 40 and under, 0.8 percent of employed men and 0.5 percent of employed women have participated (0.6 percent overall). Next we make simple comparisons between the level of RC participation for specific occupational groups and the average participation across all groups. The extent to which RC participation within any particular occupational group differs from average participation suggests that there may be difficulties associated with participation. Note that all of the comparisons between occupation-specific participation in the RC and overall participation in the RC are statistically significant at the $p < .001$ level, with the exception of women under age 41 in the food preparation and serving-related occupations, which are statistically different at the $p < .01$ level.

Summary of General Social Survey and American Community Survey Findings

As noted, the GSS highlighted three occupational groups with high shares in seven of the nine conflict variables: health care practitioners and technical occupations, protective service occupations, and transportation and material moving occupations. Apart from female transportation and material moving workers, none of these broad occupation groups reported substantially lower participation in the reserves than the average—and some, such as protective services, reported substantially more. Specific occupations in these broader occupational groups—especially health care practitioners and technical occupations—do show substantially lower participation in the RC, however. Physicians and surgeons, licensed and practical nurses, pharmacists,

Table 4.2

Reserve Participation, by Occupation, American Community Survey, 2011–2015 (Abridged)

Occupational Group	Ever in the Reserves (Men)	Ever in the Reserves (Women)	Ever in the Reserves (Age <41, Men)	Ever in the Reserves (Age <41, Women)
All	1.5%	0.5%	0.8%	0.5%
Community and social service	1.6%	0.5%	0.9%	0.3%
Computer and mathematical	1.1%	0.4%	0.8%	0.2%
Construction and extraction	1.1%	0.3%	0.6%	0.5%
Education, training, and library	1.5%	0.4%	0.7%	0.3%
Food preparation and serving related	0.8%	0.4%	0.6%	0.4%
Health care and technical	1.5%	0.6%	1.0%	0.6%
Management	2.0%	0.5%	1.0%	0.5%
Protective service	2.4%	1.0%	2.2%	1.1%
Transportation and material moving	1.6%	0.4%	0.9%	0.3%

SOURCE: U.S. Census Bureau, American Community Survey data (abridged).

and occupational therapists were among the occupations that reported substantially lower than average participation in the reserves. Among the transportation and material mover occupations, sailors and marine oilers, ship engineers, air traffic controllers and airfield operations specialists, ambulance drivers, bus drivers, and flight attendants all reported participation in the RC at rates even lower than the health care practitioners mentioned, at less than half the national average.

Broadly speaking, these specific occupations possess skills that are closely aligned with many relevant skills for the military (and particularly for some of the previously identified recruitment shortfalls from

Chapter Three). Although we are unable to verify that RC members did not enter these occupations following their reserve service, recall that we focus on individuals age 40 and under.

The GSS also identified two occupational categories with potentially high levels of conflict in six of the nine variables: community and social service occupations and construction and extraction occupations. These groups report generally similar levels of RC participation to the national average. When we look at specific occupations within these groups, clergy reported participation in the RCs at about the national average, although all other types of religious workers (e.g., directors of religious activities and education) were among those who were substantially underrepresented.

The next three occupational groups reporting the most areas of conflict in the GSS—education, training, and library occupations, food preparation and serving-related occupations, and management occupations—tended to report slightly lower levels of reserve participation (with the exception of male management-related occupations). Specific food service occupations, such as chefs, head cooks, and food preparation workers, had participation rates well below the national average. The groups reporting the lowest reserve participation in the ACS were men in arts, design, entertainment, and sports occupations;[32] men in media occupations; men and women in farming, fishing, and forestry occupations; men in legal occupations; and women in computer and mathematical occupations. It is worth noting that the arts, design, entertainment, sports, and media occupations also reported the highest reported share with irregular/on-call schedules in the GSS, as well as among the highest shares when also including split and rotating schedules.

When taken together, the descriptive results from the GSS and ACS analyses suggest (as hypothesized) that factors associated with work schedules including irregular/on-call hours, split/rotating shifts,

---

[32] It may or may not be a factor, but it might be worth studying efforts to recruit professional athletes for RC duties to explore the cultural dynamics of their service, such as perceptions of the risk of being wounded or killed or the history of athletes' military service and Vietnam-era draft dodging (see, e.g., Stefan Fatsis, "What Patriotism Means to the NFL," *Slate*, May 31, 2018).

and frequent extra hours may be hard to reconcile with reserve service. Similarly, some occupations with particularly taxing mental or physical requirements that lead workers to feel used up, especially tired, or unable to meet family demands appear to be less conducive to service in the RCs. However, there are exceptions to these general findings and cases of occupations that would appear to have high barriers to RC service that in fact have higher-than-average RC participation rates. This suggests that other factors may be driving the differences we observe, and we are unable to untangle these effects.

As such, the comparison of observed RC participation by occupational group to the national average is merely suggestive of the extent of potential conflicts with ability to participate; it should be considered further in conjunction with the reports of potential conflict. Lower-than-average participation rates in the RCs among occupation groups does not necessarily mean there are work-based barriers to participation, as there may be little skill overlap between some occupations and the skills utilized in the reserves, which could contribute to lower overall interest in participation. Similarly, higher-than-average participation in the reserves by certain occupational groups does not necessarily indicate a lack of conflicts to reserve participation. For instance, protective service occupations report substantially higher participation in the reserves than other occupational groups, but they also report among the greatest number of areas of potential conflict with participation. Thus, despite the potential barriers to participation, many people in this occupational group are able to participate. However, it is not clear how many more would be able to participate were the potential conflicts reduced.

Examples of Underleveraged Occupations and Populations

Chapter Three identified a range of occupations experiencing shortfalls across the services. In this section, we consider ways to align underleveraged occupations and populations with the military shortfalls they might help alleviate.

Aviators

Aviators posed a particular analytical challenge. On one hand, they were identified across all the services as being in high demand. On the other hand, there is a substantial recruitable population of civilian pilots, as the Federal Aviation Administration reported[33] that in 2017 there were roughly 600,000 active certified pilots in the United States; of these, about 100,000 were commercial pilots, and a further 160,000 were airline transport pilots (approximately 40 percent under age 40). Additionally, the ACS indicated that aircraft pilots and flight engineers were substantially more likely to participate in the RC than the national average. One interpretation of these data points is that military demand for pilots is so high that even high rates of participation by civilian pilots does not generate the absolute level of participation needed by the services.

One way to overcome this disparity would be to find ways for the civilian pilots to do more military duty, on average. This would require them to have adequate flexibility in their civilian work schedules to fly more often each month. This may be a challenge, as aircraft pilots and flight engineers reported one of the highest shares of work with irregular or on-call schedules in the GSS and were similarly among the most likely to report that when they worked overtime it was required by their employer, that it was hard to take time off, and that the job often interfered with family life. Clearly, pilots face strong time and location constraints to RC participation, which any effort to increase participation will need to address.

Cyber Professionals

Cyber skills were also identified as an area of shortfall across the services. Although it is difficult to measure the size of the existing civilian population with the most relevant cyber training, we do know that there are an estimated 5.2 million Americans working in potentially

---

[33] Federal Aviation Administration, "U.S. Civil Airmen Statistics," webpage, February 16, 2018.

relevant occupations, half of whom are under age 40.[34] ACS analysis indicated that many of these occupations had lower participation in the RC than the national average. This was not unexpected, given that workers in computer and mathematical occupations were most likely among the 22 occupational groups to report feeling too tired to do chores and that it was difficult to fulfill family responsibilities several times a week, according to GSS data. Web developers and information security analysts also reported much higher rates of irregular schedules and irregular, split, or rotating shift work than the average.

Health Care Professionals

The Army, Navy, and Air Force each identified health care professions broadly, as well as including a range of specialties, as needed skills. Health care practitioners and technical occupations included roughly 9.1 million civilian workers in 2017,[35] about 45 percent of them under age 40. This includes 1.1 million physicians and surgeons, 3.1 million registered nurses, 223,000 emergency medical technicians and paramedics, 122,000 occupational therapists, and 342,000 pharmacists. Overall, health care practitioners and technical occupations placed among the occupational groups with the most identified areas of conflict, placing highly on seven of the nine aspects assessed by the GSS. Thus, it is not surprising that many of the specific occupations within this category, such as physicians, surgeons, and pharmacists, were among the occupations that reported substantially lower than average participation in the RC.

---

[34] Relevant occupations include computer and information systems managers, computer and information research scientists, computer systems analysts, information security analysts, computer programmers, software developers, application and system software developers, web developers, computer support specialists, database administrators, network and computer systems administrators, and computer network architects.

[35] According to Current Population Survey estimates for 2017, sponsored jointly by the U.S. Census Bureau and BLS. See BLS, "Labor Force Statistics from the Current Population Survey," webpage, January 19, 2018a.

Chaplains

Chaplains were specifically identified by the Navy as in short supply, but our interviewees agreed that this applies to the other services as well. There are an estimated 406,000 active clergy in the United States, although only about 25 percent are under age 40. Expanding that number to include directors of religious activities and education, and all other religious workers, adds an additional 166,000 (about one-third under age 40). It is worth noting that not all of these religious workers are ordained clergy and therefore eligible for the chaplaincy. (We will return to this in our workforce constructs in Chapter Seven.) Similar to pilots, clergy did not report substantially lower participation rates in the RC than the national average—but again this does not mean they do not face constraints against greater levels of participation. Clergy reported high prevalence of irregular schedules and irregular, split, or rotating schedules. Further, many of these individuals typically work during weekends during required RC training.

Culinary Specialists

As of the time of the writing of this report, the Army offers enlistment and retention bonuses for culinary specialists, suggesting they are also needed.[36] There are more than 3 million current workers in the food preparation industry that may have closely matched skills, including cooks (2.08 million), chefs and head cooks (465,000), and first-line supervisors of food preparation and serving workers (519,000); 65 percent of these workers are under age 40. Analysis of the GSS indicated that food preparation and serving-related occupations were among the occupation groups that had the highest levels of occupational conflict—schedules were reported to be unpredictable, it was hard to take time off, and individuals reported high levels of feeling used up after work and that the job interfered with family life. Chefs and head

---

[36] The eligible specialties and bonus levels can be changed by the services whenever they feel more or less is needed to meet entry-level requirements. At the time we researched this, the other services were not offering a recruiting bonus for this specialty. See Table A.2 for the full list of bonus specialties at that time.

cooks, and food preparation workers, had RC participation well below the national average in the ACS.

Single Parents

Although they do not provide a specific skill set as a group, single parents may be particularly relevant to RC enlistments, even more so than to AC enlistment. As noted earlier, there are roughly 20 million single parents in the United States. Selected reserve members are generally older than active-duty personnel and, thus, more likely to have children at the time of enlistment (i.e., while 44.2 percent of active-duty personnel were 25 or younger in 2016, only 33 percent of Selected Reserve members were).[37] In addition, the proportion of Selected Reserves who are currently married is substantially lower than among active-duty members: 44.5 percent compared with 53.5 percent in 2016. The proportion married among Selected Reserve have also declined steadily since a recent high of 53.2 percent in 2000. Single parents are thus more prevalent among the RC than among active-duty members: 9.1 percent of all Guard and reserve personnel were single parents in 2016 (down from 9.3 in 2010), compared with 4.3 percent among the active-duty population (also down, from 5.2 percent in 2010).[38] Childcare costs can be substantial, particularly for those with preschool-age children. In 2011, the typical family spent 14 percent more on childcare than it did in 1990, although there was substantial heterogeneity across child age (families with preschool-age children increased childcare costs by 29 percent, while those with school-age children actually saw a decline in costs of 8 percent) and family structure (unmarried parents with young children saw smaller increases in expenditures than did married parents).[39] The share of all families that pay for some childcare has declined over time, from 37 percent in 1990 to 27 percent in 2011— which could suggest an unmet need and that parents are being forced

---

[37] DoD, *2016 Demographics, Profile of the Military Community*, Washington, D.C., 2016.

[38] DoD, 2016; Molly Clever and David R. Segal, "The Demographics of Military Children and Families," *The Future of Children*, Vol. 23, No. 2, Fall 2013.

[39] Chris Herbst, "The Rising Cost of Child Care in the United States: A Reassessment of the Evidence," *Economics of Education Review*, Vol. 64, 2018.

to rely more on free informal care arrangements (leaving children with relatives, for example). Single parents face additional time constraints that married parents do not, positioning them in Cell C of the conceptual model. While single parents are generally unable to join the RC unless they transfer custody of their children prior to enlisting, becoming a single parent while serving (though not formally precluding them from service after enlistment) may be an important cause of attrition from service that could potentially be attenuated if their unique needs were addressed—which will be further considered in Chapter Five.

Personal and Occupational Characteristics May Limit Reserve Component Participation Among Some Populations

This section has described an array of characteristics that may preclude individuals from service in the RCs altogether or that may present barriers to service without necessarily preventing service per se. In addition to personal characteristics identified in regulations that preclude service, we developed a conceptual model of characteristics that present barriers to service along two dimensions: location and time. We probed the concepts highlighted in this model using a variety of sources of nationally representative data, illustrating that, in some cases, individuals employed in occupations with characteristics that presented barriers to service along either of these dimensions tended to report lower levels of participation in the RC than the national average participation rate. We also identified several subpopulations, such as religious workers and health care professionals, who are in high demand from the services but who face barriers to service that could potentially be systematically reducing their participation. Similarly, we highlighted the substantial population of single parents, who could similarly face barriers to service but who also might potentially possess relevant skills of interest across the services. It is critical to note that these findings are illustrative and that the subpopulations highlighted here may not in every case face these specific barriers to participation in the RC. Nonetheless, these findings help to highlight potential barriers to RC participation that might be usefully targeted through alternative work-force constructs to raise the participation of individuals with specialty skill sets in demand in the RC.

The current chapter has highlighted the potential utility of addressing the scheduling and location barriers faced by potential RC participants. In the next chapter, we examine ways other militaries and the private sector have adapted work to provide flexible arrangements for participants, in some cases with these issues explicitly in mind.

CHAPTER FIVE

A Comparative Analysis of Reserve Component Organizational Models Across Foreign Militaries

Alongside the United States, many other countries maintain reserve forces to provide for their national defense. These RCs differ in size, shape, and relation to the active force. Certain other countries' RCs may offer insights or suggestive models for consideration by the United States, because they illustrate different ways to use reserves to complement active forces, access new parts of the population, or structure reserve service to complement civilian life.

To identify those analogous foreign RCs and models with relevant insights, RAND performed an initial broad-scope examination of all foreign militaries, exploring variables such as recruitment versus conscription, variable participation, flexible work arrangements, and experience in similar operational environments to the United States (or service alongside U.S. forces). RAND's initial screening also looked for specific shortages with these foreign militaries and whether reserve forces were used as an alternative strategy to fill these shortages. This initial research phase generated a set of 37 countries as potential case studies. Of these, 23 countries had potentially relevant models and some amount of information publicly available or easily accessible; 14 countries did not appear to have applicable or relevant models for consideration.

Within this bounded set of 23 countries with potentially relevant case studies and accessible information, RAND selected four country case studies based on similarities of force structure, societal context, and operational employment to the U.S. military, as well as the extent to which their RCs contained models that may be relevant to current problems faced by the U.S. military. These include the United Kingdom, Australia, France, and Estonia. Each case study is presented in brief below. These sections describe the force structure, societal context, and operational employment for each military; one or more relevant RC model found in each military; and the applicability of that model for the U.S. military. We did not attempt a holistic assessment of each country's RCs and their manning successes and challenges; in addition to adding considerably to the length this report, it would likely have yielded extensive, country-specific details that simply would not have been applicable to the challenges facing other countries, including the United States.

United Kingdom

Background on the British Reserve Components

The British armed forces include a total full-time strength of 153,294 service members, divided between 85,992 Army soldiers, 34,057 airmen in the Royal Air Force, and 33,245 personnel in the Royal Navy and Royal Marines.[1] The British military reserves include 85,716 personnel, split between 61,151 Reserve Land Forces, 11,661 Reserve Air Forces, and 12,404 Reserve Naval and Marine Forces.[2] The ACs and RCs of the military add up to 239,010 personnel, or 0.3 percent of the total UK population of 65,105,246.[3] Like the United States, the British military currently enlists volunteers only, although it used conscription during World War I (through the Military Service Act of 1916) and World War II (through the National Service [Armed Forces] Act of 1939). The British military predominantly draws on British citizens (subjects) to fill its ranks; however, it maintains historical units

---

[1]   UK Ministry of Defence, "UK Armed Forces Quarterly Service Personnel Statistics, 1 July 2018," August 23, 2018.

[2]   UK Ministry of Defence, 2018.

[3]   Central Intelligence Agency, "United Kingdom," *The World Fact Book*, last updated October 23, 2018.

that recruit from Nepal (Gurkhas), Northern Ireland, and other Commonwealth communities across the historical British Empire.[4]

Historically, the British military has used its reserves in time of war or national emergency or as necessary to augment its active forces. Recently, and in parallel with the U.S. military, the British military has sought to operationalize its RCs and leverage their comparative advantage as forces grounded in the civilian community and workforce. These changes have unfolded over a series of British strategy documents, including the 2010 Strategic Defence and Security Review (SDSR), the 2010 Future Force 2020 vision, the 2012 *Army 2020* force structure review, and the 2015 *National Security Strategy and Strategic Defence Review.*[5] Greater operational use of the reserves evolved alongside decreases in the number of active personnel, which drove new roles for the British reserves, who would be called upon to provide "additional capacity as well as certain specialists whom it would not be practical or cost effective to maintain in the regular forces."[6]

Notably, in shifting to a "Whole Force Approach" that integrates active and reserve manpower, the British military explicitly states that "it is more effective and efficient to draw on specialist capacities from dedicated providers and the civilian labour force, rather than to develop such skills indigenously within the regular force structure."[7] Among the specialty areas where this is most true are medicine, linguistics, and cyber security, where the British military intends to heavily leverage its RCs to access expertise that cannot be efficiently or effectively recruited and maintained by the active force.

---

[4] Central Intelligence Agency, "United Kingdom."

[5] Timothy Edmunds, Antonia Dawes, Paul Higate, K. Neil Jenkings, and Rachel Woodward, "Reserve Forces and the Transformation of British Military Organisation: Soldiers, Citizens and Society," *Defence Studies*, Vol. 16, No. 2, 2016, p. 119.

[6] UK Ministry of Defence, *UK Reserve Forces and Cadets Strengths, 1 April 2015*, London, June 18, 2015, p. 4.

[7] Edmunds et al., 2016, p. 121.

The Sponsored Reserve Concept

In 1996, Britain legislatively created a type of reserve force structure known as the sponsored reserves. The 1996 Reserve Forces Act authorized the Ministry of Defence to enter into contractual relationships with private companies providing services to the UK military and treat the workforces of those services contractors as sponsored reserves in the event that they accompany British forces on deployments into hostile areas. These service contractors (and their employees) provide a range of capabilities to the Ministry of Defence, including transportation and engineering that are not cost effective to maintain within either the regular or volunteer reserve force. They can be mobilized to serve in military uniform to accompany British forces on operational deployments into nonbenign areas to continue to provide a capability, where it is not appropriate to risk the use of civilian contractors. In doing so, they are subject to the same military regulations, including discipline, as other UK armed forces personnel. However, during these deployments, the contractors continue to work as employees of their private firms, drawing pay and benefits set by their employers (who, in turn, bill the ministry under the terms of their respective government contracts). The program currently includes more than 2,000 personnel, or approximately 6.5 percent of the total British military reserve, spread across more than ten service contractors. Although the program remains limited to certain niche specialties within the British military and its contractor base, it has been utilized as required on a frequent basis to support Ministry of Defence missions, including recent operational deployments.[8]

The primary value of the sponsored reserve program lies in its cost-effective use of capability that resides primarily in the private sector, as well as in how sponsored reservists are fully integrated into the armed forces through their home-base roles. Sponsored reserves are reportedly cheaper than other British military personnel and more cost effective for meeting certain types of capability requirements in the

---

[8] Interview with UK Ministry of Defence personnel.

areas of transportation, maintenance, and sustainment.[9] The reason is that the Ministry of Defence does not own the wider infrastructure costs and long-term liabilities for these personnel, such as housing and pension costs. The secondary value is their enhanced survivability by virtue of being trained and equipped as military reservists, which helps prepare them for the rigors and dangers of service in nonbenign areas. Relatedly, by deploying as support personnel instead of civilian contractors, British sponsored reserves are considered combatants under the Geneva Conventions, clarifying any potential legal status questions that might surround the operational employment of civilian contractor personnel accompanying military forces in an armed conflict.[10]

Notably, the United Kingdom does not use sponsored reserves (nor civilian contractors) to perform security or combat-related functions in areas of active conflict. Current sponsored reserve units include the Royal Fleet Auxiliary, which provides the support vessels for the British Navy, and was the operating model that was partly used to create the sponsored reserve program in the mid-1990s. Other sponsored reserve units exist within the maritime and aviation fields, primarily to maintain major systems. It also uses sponsored reserves for smaller niche capabilities, such as its mobile meteorological unit.[11] All of these units operate in ways similar to other types of reserves and are indistinguishable in their uniforms and basic equipment. There are slight differences of organization and command and control, but these are largely managed on a case-by-case basis depending on the type of sponsored reserve unit and its nexus to the supported unit or element. To date, the British military has not experienced difficulty recruiting for its sponsored reserves. Across all of these sponsored reserve units, most personnel are former active-duty personnel who apply for jobs with a sponsored reserve obligation and volunteer to continue service as a sponsored reserve as part of their civilian employment.

---

[9] Interview with UK Ministry of Defence personnel.

[10] Interview with UK Ministry of Defence personnel.

[11] UK Ministry of Defence, 2015, p. 8.

Applicability to the U.S. Context

The British military uses the sponsored reserve concept primarily to manage the legal status of support personnel who accompany military forces into areas of armed conflict. The U.S. military uses a blended Total Force of active, reserve, civilian, and contractor personnel to accomplish these functions in the same environments.[12] Although legal questions have been raised regarding U.S. government use of civilian contractor personnel in conflict zones and for combat-related functions, such as protective security, the government has decided that such operational employment of civilians or contractors is appropriate as a matter of policy. Therefore, the British sponsored reserve concept may not be directly transferable to the U.S. military.

However, the British model may help the U.S. military solve other problems should the latter continue to leverage operational support contractors for future military operations. A number of questions have emerged regarding the adequacy of health care, disability compensation, and related programs for support contractors.[13] Deploying these personnel with the status of military reservists may solve this problem by changing their status to military personnel, removing them from insurance schemes (such as that authorized by the Defense Base Act), and entitling them to long-term VA care and support. Deploying operational support personnel as military reservists may also address questions of accountability for conduct while deployed and make them fully subject to the U.S. military justice system. Furthermore, deploying operational support personnel in a military status could also reduce the litigation risk associated with adverse battlefield events, such as environmental exposures, accidental deaths, or combat deaths, which have resulted in substantial amounts of tort litigation.[14] Under current federal law, service members cannot sue the federal government in tort

---

[12] Molly Dunigan, "The Future of US Military Contracting: Current Trends and Future Implications," *International Journal*, Vol. 69, No. 4, 2014.

[13] Dunigan, Farmer, et al., 2013.

[14] See, for example, *Harris v Kellogg, Brown & Root Services, Inc.*, 724 F.3d 458 (3d Cir. 2013); *Al Shimari v CACI Premier Tech., Inc.*, 840 F.3d 147 (4th Cir. 2016); In re KBR, Inc., Burn Pit Litig. ("Burn Pit III"), 744 F.3d 326 (4th Cir. 2014).

for injuries or illnesses sustained on duty. A sponsored reserve concept, as applied to the workforces of operational support contractors, may eliminate such litigation, substituting VA access for the right to sue for torts, such as wrongful death or negligence.

From a total manpower perspective, such a program creates a net gain in one of two ways. First, the individuals within it might end up serving in situations where they would not be able to serve as contractors. For example, if there were contingency operations that took place too rapidly for a contract to be put in place, or if there were other external reasons why DoD chose not to use contractors, individuals already "on the rolls" as sponsored reservists might be mobilized to perform their identified tasks and missions. Second, there may be individuals who would not have joined either the military or their contractor without the appeal of both careers.

Australia

Background on the Australian Reserve Components

Australia maintains an active force of 59,200 personnel, including 30,700 soldiers in the Army, 14,400 Air Force personnel, and 14,100 personnel in the Australian Navy.[15] (Australia does not have a naval infantry branch equivalent to the U.S. Marine Corps.) Alongside this active force, Australia maintains a reserve component that consists of 14,800 Army reservists, 3,100 Air Force reservists, and 1,850 Navy reservists. The active and reserve components of Australia's military add up to 78,600 personnel, or 0.3 percent of the total Australian population of 23,232,413.[16] Although Australia has historically relied on conscription during both wartime and peacetime, it moved to an all-volunteer, recruited force in 1972, much as the United States did at the

---

[15] Jane's, "Australia: Armed Forces," September 8, 2018c.

[16] Central Intelligence Agency, "Australia," *The World Factbook*, last updated November 6, 2018.

end of the Vietnam War.[17] Australia now recruits military personnel from its youth population, with the majority drawn from working- and middle-class families. However, Australian political and military leaders have expressed concern in recent years about the country's ability to produce sufficient numbers of recruits and about the Australian military's ability to attract this talent.[18]

Australia uses its RC as a pool of individuals who may volunteer for specific missions, in contrast to the U.S. and British practice of mobilizing units or individuals for active duty at the discretion of the national or military leadership. According to an Australia Strategic Policy Institute study, "Since World War II, the deployment of Australia's Reserve forces at home and overseas has relied exclusively on individuals volunteering to serve. . . . [N]o individual or reserve unit has been legally required to serve."[19] Australia's government does retain the legal authority to mobilize reservists or reserve units. This is accomplished using a legal mechanism called a "call out" under Australian law, which requires a specific declaration by the Australian governor general (comparable to a U.S. declaration of national emergency that is required under 10 U.S.C. 12304).[20] However, this procedure has not been used for the reserves since at least the Vietnam War. Recent Australian deployments have been supported by either the active force or reserve personnel who have volunteered for deployment.

The Australian Service Category Model

The Australian Defence Force (ADF) Total Workforce Model (TWM) enables it to flexibly mobilize, assign, and utilize reservists across the

---

[17] Gary Brown, *Military Conscription: Issues for Australia,* Australian Parliament, Foreign Affairs, Defence and Trade Group, October 12, 1999; Australian National War Memorial, "Conscription," *Memorial Encyclopedia*, last updated October 23, 2018; National Archives of Australia, "Universal Military Training in Australia, 1911–29," Fact Sheet No. 160, undated.

[18] See, for example, Mark Eggleton, "Australian Armed Forces Face Demographic Crisis," *Financial Review,* March 16, 2015; see also Brown, 1999.

[19] Andrew Davies and Hugh Smith, *Stepping Up: Part-Time Forces and ADF Capability*, Australia Strategic Policy Institute, Strategic Insights No. 44, November 2008, p. 3.

[20] Davies and Smith, 2008, p. 4.

force. The TWM is a triservice framework designed to provide maximum flexibility to both the service and the individual. This flexibility responds well to operational needs from Australian military units, as well as personal needs from individual reservists seeking a balance between their personal circumstances and service needs.[21] The TWM encourages mobility between the full-time and part-time components of the ADF and "enhances the services' ability to draw efficiently upon different workforce mixes to meet capability demand."[22] The model also enables the Australian military to retain personnel who might otherwise leave the military altogether if it were not for more flexible assignment policies. This, in turn, helps the Australian military maintain higher-quality personnel in its RCs who agree to make themselves available on a flexible basis to the force.[23]

The heart of the TWM is its Service Spectrum consisting of seven distinct Service Categories (SERCATs) and four Service Operations (SERVOPs). The SERCATs define the individual obligations of service, liability for call out (mobilization), and other parameters of military service, such a compensation and tax liability. As described by the Australian DoD, "a SERCAT groups members into like service and duty arrangements that share mutual obligations and conditions of service. All members are categorised in a single SERCAT at all times."[24] SERCATs 1 through 5 are for reserve personnel; SERCATs 6 and 7 exist for active personnel. By design, the SERCAT model makes it easy for Australian reserve members to move between categories of service, with as little friction as possible, managing their careers as they choose to complement their civilian employment and family lives.[25]

SERVOPs are another kind of Australian personnel authority that exist alongside SERCATs, providing a specific package of assign-

---

[21] Australian Department of Defence, "ADF Total Workforce Model," webpage, undated(a).

[22] Australian Department of Defence, *ADF Total Workforce Model, Frequently Asked Questions*, version 3.0, Canberra, 2018, p. 5.

[23] Australian Department of Defence, undated(a).

[24] Australian Department of Defence, "The Service Spectrum," undated(b).

[25] Australian Department of Defence, undated(a).

ment, mobilization, compensation, and administrative parameters that are tailored to specific unit or mission requirements. An example is SERVOP C: "Reserve members serving in SERCATs 3, 4 or 5 who are rendering Continuous Full-Time Service."[26] Within this package, reservists serve for a defined period of time on active duty and receive similar compensation and benefits as active personnel (those in SERCAT 7) for this service.[27]

The ADF's experience with this model is relatively new. Each of the ADF's services were charged with implementing the TWM in late 2016, with discretion to modify the schedules as may be necessary to fit operational needs.[28] In announcing the program in 2016, service leaders pledged to make it a centerpiece of their efforts to improve quality of life for service members while also improving readiness and operational efficiency.[29] However, it is too early to fully evaluate the operational experience of this model.

Applicability to the U.S. Context

The Australian TWM approach resembles some of what already exists in statute, regulation, and practice for the U.S. military. Although different from a legal standpoint, the SERCATs are functionally similar to the mobilization authorities found in 10 U.S.C., which similarly define the legal predicate, operational conditions, pay, benefits, and status of RC personnel.[30] Likewise, there is a functional similarity between the Australian SERVOP concept and the package of service, pay, and benefits contained in a particular mobilization or deployment order.

At present, the Australian reserve system has fewer categories than the U.S. system. This difference may shrink if the U.S. Con-

---

[26] Australian Department of Defence, undated(b).

[27] Australian Department of Defence, undated(b).

[28] Australian Department of Defence, undated(a).

[29] Harley Dennett, "Giving a Commitment Choice with the 'Service Spectrum,'" *The Mandarin*, April 20, 2018.

[30] See, for example, 10 U.S.C. 12304, Selected Reserve and Certain Individual Ready Reserve Members; Order to Active Duty Other Than During War or National Emergency, January 12, 2018.

gress continues to legislate the reform of RC duty statuses.[31] A simpler, more clearly defined system of duty statuses could simplify the process as individual reservists navigate between regular positions and temporary assignments. Such a system might also complement initiatives like the continuum of service concept (advanced by the U.S. Navy and Army in recent years), designed to facilitate movement of individuals between active and reserve service over the course of their careers.[32] As the Australian model matures and the Australian military gains practical experience through its implementation and use, there may be lessons to be learned by the U.S. military as it considers similar policy structures for the mobilization, assignment, and utilization of RC units and personnel.

Another vector of applicability may be the way the Australian military leverages the SERCAT system to create an intermediate level of service between traditional drilling reservists (who serve approximately 40 days per year) and full-time service. This intermediate level of service is comprised of episodic SERCAT tours in an active status, comprising between 120 and 150 days per year, and often includes a deployment or other finite type of service. As the U.S. military adjusts its duty-status paradigm for reservists and seeks to utilize reservists for something in between traditional drill and full-time service, the Australian model may provide insights into how such a system could work in practice.

France

Background on the French Reserve Components

France has an active defense force that includes approximately 301,324 personnel, divided between 112,500 soldiers in the French Army,

---

[31] Arnold L. Punaro, Chair, Reserve Forces Policy Board, "Report of the Reserve Forces Policy Board on Reserve Component (RC) Duty Status Reform," memorandum, July 16, 2013b.

[32] Note that because both duty-status reform and continuum of service are already under way, we did not include them in our evaluation of alternative workforce constructs.

35,600 sailors in the French Navy, 41,200 airmen in the French Air Force, 95,200 national police in the Gendarmerie, and several smaller formations, such as the French Defence Health Service.[33] Alongside these active forces, France maintains reserves of approximately 70,000, including 30,000 in the Gendarmerie, 18,800 in the Army reserves, 5,200 in the Navy reserves, and 4,800 in the Air Force reserves.[34] The AC and RC of the military add up to 371,324 personnel, or 0.5 percent of the total French population of 67,364,357.[35] France fills its active forces through a voluntary service model, drawing on men and women aged 18–25 who serve a minimum of one year. There is a slightly broader window of enlistment for the reserves, which allow new enlistees aged 17–35, with service permitted in some specialties until age 72.[36] While serving actively, French reserve service members receive full compensation equal to that of active personnel.[37] In addition, many of those who have jobs continue to be paid for at least some of their time away from work.[38] To encourage enrollment, the French military also offers bonuses to help pay for school and other professional training, as well as to subsidize the acquisition of a driver's license, which is very expensive in France. There is also a bonus for reservists who reenlist.[39]

---

[33] Haut Comité d'Evaluation de la Condition Militaire [Military Status Assessment Committee], *Revue Annuelle de la Condition Militaire* [*Annual Review of the Status of the Military*], Paris, 2017, p. 21.

[34] Gaëtan Poncelin de Raucourt, *Compte rendu: Commission de la défense nationale et des forces armées* [*Report: National Defense and Armed Forces Commission*], Paris: French National Assembly, 2017, p. 10.

[35] Central Intelligence Agency, "France," *The World Factbook*, last updated February 5, 2019.

[36] Direction de l'information légale et administrative (Premier ministre) [Directorate of Legal and Administrative Information, Office of the Prime Minister], "Réserve opérationnelle" ["Operational Reserve"], August 23, 2018.

[37] Direction de l'information légale et administrative (Premier ministre), 2018.

[38] Ministère des Armées [French Ministry of the Armed Forces], Rapport d'évaluation de la réserve militaire *2016* [Military Reserve Evaluation Report, *2016*], Paris, 2017, p. 44.

[39] Direction de l'information légale et administrative (Premier ministre), 2018.

After the Cold War, the French maintained reserves largely to retain access to individuals with specific skills in numbers that the active services could not justify supporting. Today, the French military, for example, is working on cultivating reserve cyber cadres that can draw on the skills found in the civilian workplace. In more recent years, however, the growth of the French military's operational requirements (in particular a large homeland defense mission, Operation Sentinelle, which began in 2015) have motivated France to grow its RCs to alleviate the general burden on active forces. A small number of reservists are deployed for overseas operations, but the vast majority serve in metropolitan France.[40] Mostly this means Sentinelle, but others guard military installations or support various government agencies dealing with disasters such as forest fires. The French military also uses reservists with highly specialized skills for a number of relatively small support services. These include the French military's defense technology development and procurement agency, the Direction Générale de l'Armement, the military commissary, the fuel service, and the health service.[41]

Reservists often are organized as companies that are assigned to active battalions, with a few individuals assigned to headquarters functions. The active and reserve companies train together and are deployed together. There is at least one regiment (the 24th Infantry Regiment) comprised entirely of reservists. Based in Paris, the 24th can be thought of as akin to a local U.S. National Guard unit.[42]

---

[40] In 2016, only 49 reservists deployed. In contrast, an average of 7,200 French Army personnel deployed each month. See Haut Comité d'Evaluation de la Condition Militaire, 2017, p. 41.

[41] Association des Officers de Réserve des Corps de l'Armement, "La réserve militaire DGA," AORCA: Association des Officiers de Réserve des Corps de l'Armement (blog), undated; Ministère des Armées, "La Garde Nationale," undated(a); Ministère des Armées, "Réserviste au Service de santé des armées (SSA): une vie pleine d'aventure," La réserve militaire, composante de la garde nationale, July 18, 2016.

[42] Armée de Terre, Création du bataillon de réserve Ile de France—*24*e régiment d'infanterie à Vincennes, July 2, 2013.

The Garde Nationale

In October 2016, France rebranded its various reserves as the Garde Nationale, which evokes a historical tradition of middle-class volunteer formations created first during the French Revolution under the Marquis de Lafayette.[43] They accompanied the name change with a public communications effort to publicize the institution and encourage recruitment; the Garde's commander in testimony to the French Senate in December 2017 asserted that the publicity campaign has helped both recruitment and public support, especially from those who employ reservists as civilians.[44] The Garde Nationale includes two basic types of reservists: the Citizens Defense and Security Reserve and the Operational Reserve. The Citizens Reserve exists to foster the "spirit of defense" in the civilian world and basically act as ambassadors charged with fostering positive civil-military relations.[45] The Operational Reserve itself has two tiers: direct volunteers referred to as ESR reservists (*engagement à servir dans la reserve*, or "engagement to serve in the reserve") and outgoing servicemen and women who remain liable for activation for five years after leaving.[46] The vast majority of reservists who serve on active duty are ESR reservists; the recently separated servicemen and women by and large are not tapped for duty and effectively represent a strategic reserve. ESR are committed to 30 days of active service but are liable to be called up for up to 210 days if needed.[47] The average in 2017 for all members of the Operational Reserve was 35 days, up from previous years despite growing numbers of ESR reservists (a reflection of the burden of Sentinelle).[48] The average annual service for Army reservists is substantially higher than

---

[43] Faustine Vincent, "La garde nationale est un simple label," *Le Monde*, October 12, 2016.

[44] Poncelin de Raucourt, 2017, p. 4.

[45] Haut Comité d'Evaluation de la Condition Militaire, 2017, p. 31.

[46] Haut Comité d'Evaluation de la Condition Militaire, 2017, p. 31.

[47] Haut Comité d'Evaluation de la Condition Militaire, 2017, p. 33.

[48] Poncelin de Raucourt, 2017, p. 6. See also Haut Comité d'Evaluation de la Condition Militaire, 2017, p. 34.

for the other services: 40 days for officers in 2016, 44 for noncommissioned officers (NCOs), and 28 for enlisted soldiers.[49]

The French military places a value on recruiting as broadly as possible from French society and hopes by means of the Garde Nationale to reach categories of citizens that might not otherwise serve. They may also see it as a way to exploit what they see as a surge in interest in serving among young people generated by the terrorist attacks of 2015—what the French call the "Bataclan Effect" after the site of one of the attacks."[50]

The French military generally expresses satisfaction with both the number of recruits and the kinds of recruits, although data that might validate this impression are hard to come by. According to a general interviewed by a local newspaper, fully 70 percent of reservist recruits have no military background.[51] He also noted that 20 percent were students, while 15 percent were retirees, and 5 percent were unemployed. Also, proportionately more women serve in the Garde: in 2017, 20 percent of the Garde are female, compared with about 15 percent in the active force.[52] In July 2018, a female general of the Gendarmerie, Anne Fougerat, was appointed Garde commander.

French Coordination with Reservists' Employers

An additional aspect of the French approach to cultivating its RC is its practice of entering into voluntary contracts with employers. Businesses that employ reservists sign accords with the French government in which they agree to provide those employees certain protections and benefits: a certain number of days off, a certain number of days off that are paid, an agreed-upon minimum amount of warning employees have to give before departing, and so on. In exchange, the busi-

---

[49] Haut Comité d'Evaluation de la Condition Militaire, 2017, p. 33.

[50] Nathalie Guibert, "Ruée des jeunes Français vers les armées," *Le Monde*, November 20, 2015.

[51] "70% des réservistes sans passé militaire" ["70% of Reservists Do Not Have a Military History"], *Le Courrier de l'ouest*, April 30, 2018.

[52] Poncelin de Raucourt, 2017, p. 3.

ness obtains certain tax advantages and can advertise its cooperation with the military. The French military is also counting on employers to value the skills and other intangible benefits that having employees with military experience might bring them. Intriguingly, the French military has found small businesses to be far more interested in signing such contracts than large ones. According to an Armed Forces Ministry report, in 2016, 38 percent of the signatories were businesses with fewer than 10 employees, while only 5 percent were businesses with more than 5,000 employees.[53]

Applicability of the French Model to the U.S. Context

Two areas stand out in which the French model might apply to the U.S. RCs. First, the domestic focus of the Garde Nationale and its separation from the strategic reserve of prior regular service members provides a contrast from the way the U.S. National Guard is seen as both a homeland security force and an operational force ready for major combat operations overseas.

Second, we noted how the private sector is actively recruited to formally support reservist employees in contrast to the U.S. tradition of requiring support through legislation and enforcing it through legal action. The question would be whether either of these different approaches could be adapted to the different military posture and utilization rate of U.S. RCs. Because this is, in the French case, not a structural change in the terms of service, we did not identify a parallel construct to advance in Chapter Seven.

Estonia

Background on the Estonian Reserve Components

Estonia offers a different strategic model from the countries already discussed. The Eesti Kaitsevägi, or Estonian Defense Forces (EDF), are divided into four branches: Army (Maavägi), Navy (Merevägi), Air

---

[53] Ministère des Armées, 2017, p. 43.

Force (Ohuvagi), and reserves (Kaitseliit).[54] The bulk of the EDF's capability and capacity resides in its reserves. "The Estonian Land Forces is [sic] not capable of many high-intensity warfare tasks,"[55] according to Jane's, but "its infantry forces are well-trained and equipped for small-unit conventional defensive operations."[56] To fill its ranks, Estonia relies on compulsory service of eight months for all male citizens, followed by a reserve commitment that lasts until age 60.[57] Women may volunteer to join Estonia's military as well. Estonians who rise to the rank of sergeant or choose to serve as officers serve for longer durations than their initial eight-month conscription obligation.[58] The EDF is currently comprised of approximately 7,100 active personnel, split between the Army (6,400), Air Force (400), and Navy (300).[59] Within the Army, approximately half of Estonian soldiers are professional sergeants or officers, and half are conscripts serving their first enlistment.[60]

The EDF maintains a substantial reserve of 85,000 personnel, which is nearly 12 times the size of the active force. The active and reserve components of the military add up to 92,100 personnel, or 7.4 percent of the total Estonian population of 1,244,288.[61] These reserves include an active reserve of approximately 60,000 EDF members, who conduct periodic training, consisting of one- to three-week exercises approximately every three years, to maintain military proficiency.[62] The EDF reserves include a segment of 21,000 EDF reserve members

---

[54] Central Intelligence Agency, "Estonia," *The World Factbook*, last updated October 23, 2018.

[55] Jane's, "Estonia—Army," April 11, 2018a.

[56] Jane's, 2018a.

[57] Estonian Military Service Act, June 12, 2012.

[58] Jane's, 2018a.

[59] Jane's, 2018a; Jane's, "Estonia—Air Force," December 29, 2017b; Jane's, "Estonia—Navy," July 24, 2017a.

[60] Jane's, 2018a.

[61] Central Intelligence Agency, "Estonia."

[62] Email interview with EDF officer.

who are organized for rapid reaction.[63] The remaining 25,000 members of the EDF reserves belong to the Estonian Defense League (EDL), as depicted in Figure 5.1.[64] This is a broad category of reserve membership that includes traditional reserve volunteers serving less actively, as well as honorary members, junior members serving in youth organizations, and others, including foreign citizens accorded honorary membership.[65] In general, continuing membership in the EDF is driven by voluntary participation: "Participation is ensured via a moral rather than regulatory means."[66] The EDL constitutes the alternative model appropriate for consideration in this study.

The Estonian Defense League

The EDL was established at the end of World War I as a voluntary self-defense force that is "organised in accordance with military principles, possesses weapons and holds exercises of a military nature."[67] EDL members keep weapons and equipment at their homes and mostly comprise light infantry who can defend the country against foreign invasion or augment the EDF as may be necessary.[68] In addition to providing a traditional reserve for conventional forces, the EDL provides a means for interested Estonians to render niche services, such as cyber, to the military in a part-time capacity. Such capabilities are "engaged on the twofold condition that the particular State authority is unable to complete the tasks (in a timely fashion) and that there are no other means for completing the task at hand."[69] The EDL's structural units are the EDL Headquarters, 15 geographic districts scattered through-

---

[63] Jane's, 2018a.

[64] Jane's, 2018a.

[65] Kaitseliit, "Members of the Estonian Defence League," webpage, last updated November 16, 2018.

[66] Kaska, Osula, and Stinissen, 2013, pp. 37–38.

[67] Estonian Defence League Act, February 28, 2013, Section 2(1).

[68] Andrus Padar, Commander, Cyber Defense Unit, "Estonian Defence League Cyber Defence Unit," presentation, July 2017.

[69] Kaska, Osula, and Stinissen, 2013, p. 38.

Figure 5.1
Estonian National Defense Organization

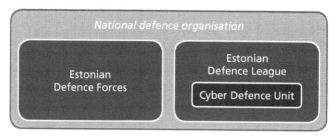

SOURCE: NATO Cooperative Cyber Defence Center of Estonia (CCDCOE), 2013, p. 10.

out Estonia, the EDL School, the Cyber Defense Unit (CDU), the women's defense organization, and youth organizations.[70] The leader of each structural unit reports to the EDL commander.[71] Unlike members of the EDF's more traditional reserves, EDL members generally are not compensated for their service but may request partial compensation for some types of service or reimbursement for costs incurred as part of their military duties.[72] EDL activities are funded by a mixture of sources including government expenditures, membership fees, private donations or sponsorship, and contractual revenue from services rendered.[73]

The EDL's CDU is "an innovative model for the involvement of volunteers in national cyber defence,"[74] according to one analyst. The CDU emerged in the aftermath of the 2007 cyber attacks against Estonian information infrastructures, thought to be orchestrated by Russia or its confederates, and was institutionalized as part of the EDL struc-

---

[70] Estonian Defence League Act, 2013, Section 9(1).

[71] Estonian Defence League Act, 2013, Section 9(2).

[72] Kaska, Osula, and Stinissen, 2013, p. 17.

[73] Kaska, Osula, and Stinissen, 2013, p. 27.

[74] Kaska, Osula, and Stinissen, 2013, p. 5.

ture in 2011.[75] In 2013, the Estonian government further codified the CDU's place in the Estonian defense structure, providing it with a legally established objective and a framework for structure, management, membership, and functioning" in the 2013 Estonian Defence League Act.[76] This statute directed the CDU to conduct "capability building and operations" for Estonia's private sector and also to provide specialized cyber defense expertise to the nation through training and exercises, peacetime technical assistance, and cybersecurity assistance during emergencies or crises.[77] To accomplish these missions, the CDU maintains personnel with broad expertise in cyber specialties who are willing to contribute to national defense on an as-needed, volunteer basis.[78] In addition to having professional expertise, CDU applicants must meet statutory age and citizenship requirements, be sufficiently healthy, and successfully pass a background check.

Unlike traditional reserves, who serve or train according to a regular schedule, CDU members (like other EDL members) serve as needed, often in a purely voluntary capacity.[79] Although the Estonian military does not disclose a great deal of detail regarding CDU activities, public reporting indicates that CDU personnel have participated in several national and NATO exercises (such as the annual NATO CCDCOE Locked Shields exercise), as well as overall defense and crisis management exercises (such as the EDF's annual Spring Storm exercise) to practice and refine cyber defense skills and information exchange procedures. The CDU has also participated in training and awareness events inside Estonia, such as cyber defense seminars for governmen-

---

[75] Kaska, Osula, and Stinissen, 2013, p. 5; see also Christian Czosseck, Rain Ottis, and Anna-Maria Talihärm, *Estonia After the 2007 Cyber Attacks: Legal, Strategic, and Organisational Challenges in Cyber Security*, Tallinn, Estonia: Cooperative Cyber Defense Centre of Excellence, undated; and Emily Tamkin, "10 Years After the Landmark Attack on Estonia, Is the World Better Prepared for Cyber Threats?" *Foreign Policy*, April 27, 2017.

[76] Kaska, Osula, and Stinissen, 2013, p. 5.

[77] Kaska, Osula, and Stinissen, 2013, p. 32.

[78] Katseliit, "Estonian Defence League Cyber Unit," webpage, last updated November 10, 2018.

[79] Kaska, Osula, and Stinissen, 2013, p. 33.

tal institutions.[80] CDU personnel have also "supported various public and private sector entities, such as supplying malware screening solutions for public school computers and assisting with the installation and security testing of the national electronic voting system."[81] And on order, as may be necessary for a future emergency, the CDU can "form and send CPT teams in case of cyber emergency."[82]

The Estonian experience with the EDL and cyber has been mixed. At an individual level, the model does enable former service members or others to contribute expertise and serve their country, particularly those who cannot join the traditional armed forces. The EDL model also gives the government tremendous flexibility and efficiency, as well as the ability to recruit and retain talent that would not otherwise be accessible to the military or able to be retained indefinitely by the military.[83] These values matter given the size of the Estonian military relative to the threat. Private-sector employers also benefit from having EDL cyber expertise in their ranks.[84] However, there have also been concerns raised within Estonia about the sustainability of the model given its loose, informal, voluntary nature. Legal questions have emerged regarding CDU's potential access to confidential information and the legal status of members during an international armed conflict.[85] And there have been private-sector concerns expressed about the availability of CDU members during a national crisis, when their employers may need them as badly as the national government, particularly if their civilian job involves the delivery of critical infrastructure services.[86]

---

[80] Kaska, Osula, and Stinissen, 2013, p. 22.

[81] Kaska, Osula, and Stinissen, 2013, pp. 22–23.

[82] Padar, 2017.

[83] Information from CDU founding member Rain Ottis cited in Monica M. Ruiz, "Is Estonia's Approach to Cyber Defense Feasible in the United States?" *War on the Rocks*, January 9, 2018.

[84] Information from CDU founding member Rain Ottis cited in Ruiz, 2018.

[85] Kaska, Osula, and Stinissen, 2013, p. 6.

[86] Kaska, Osula, and Stinissen, 2013, p. 37.

Applicability of the Estonian Model to the U.S. Context

There is clearly a need in the U.S. military for additional cyber personnel, both in the AC and RC. The Estonian CDU model illustrates one potential way to fill this need: through the creation of a less military, less formal, more permeable type of military service that would enable the U.S. military to enlist cyber operators who want to serve but do not want to serve under the traditional parameters of military service.[87]

A potential option for the creation of such an informal cyber reserve would be to create something like the CDU at the state level, as part of state militia or state military reserve organizations.[88] These organizations report to their respective governors and serve entirely in a state capacity, without the possibility of federalization or overseas deployment.[89] Such organizations could participate in military training or exercises as members of their state reserves and also participate in domestic law enforcement operations, based on their legal status under 32 U.S.C. Their compensation, rank structure, and conditions of service could be tailored to their missions and organizational requirements, on a case-by-case and state-by-state basis.

The Wounded Warriors workforce construct in Chapter Seven offers a similar model focusing on a less military reserve force. That construct is both narrower than the Estonian CDU in that it focuses on the veteran population, not all potential cyber warriors, and broader, in that cyber is only one of the possible missions for this force.

---

[87] Ruiz, 2018.

[88] California Military Department, "California State Military Reserve," webpage, undated; Maryland Defense Force, homepage, undated(a).

[89] See 32 U.S.C. 109(c), which states,

> In addition to its National Guard, if any, a State, the Commonwealth of Puerto Rico, the District of Columbia, Guam, or the Virgin Islands may, as provided by its laws, organize and maintain defense forces. A defense force established under this section may be used within the jurisdiction concerned, as its chief executive (or commanding general in the case of the District of Columbia) considers necessary, but it may not be called, ordered, or drafted into the armed forces.

CHAPTER SIX

Innovative Employment Models in Other U.S. Public Organizations and the Private Sector

In addition to reviewing innovative models to access human capital used by foreign militaries, we explored models in use in other U.S. public organizations, including nondefense federal agencies and the private sector. This review allowed us to capture a broader array of alternative work arrangements that may merit DoD's consideration. Moreover, it complements our discussion of international military models by focusing on nonmilitary models that are being used to access pools of talent in the United States.

This chapter begins by providing an overview of the state of the data and literature on nonstandard work arrangements, which we broadly define as anything other than full-time, permanent work in a fixed employer location. We describe a number of specific types of models that are in use in the private sector, elsewhere in the federal government, and in state and local governments, in many cases providing illustrative examples. The chapter concludes by discussing the potential to apply these alternative workforce models to the RCs.

Overview of Nonstandard Work Arrangements

The nature of work in the United States is undergoing a fundamental change, as full-time, long-term employment relationships give way to a "nation of freelancers" and "gig economy" workers who bounce from job to job on a project-by-project basis or who patch together a full-time job from a collection of part-time gigs. However, existing data are inconclusive regarding how widespread these arrangements have become.[1]

In fact, at the time of this writing, the most recent Contingent Worker Survey (CWS), released by BLS in June 2018, found that the share of U.S. workers in alternative work arrangements, which BLS defines to include temporary agency help workers, on-call workers, contract workers, and independent contractors or freelancers, was roughly the same in 2017 as it was when the survey was last fielded more than a decade ago in 2005 (10.1 percent in 2017 versus 10.7 percent in 2005).[2] The BLS finding differs from that of the economists Lawrence Katz and Alan Krueger, who used the RAND ALP in 2015 to estimate that 15.8 percent of workers were in alternative work arrangements (applying the same definition as the BLS).[3] Notably, considering the widespread attention they receive, Katz and Krueger found that just 0.5 percent of workers provided services through online platforms.

A constraint of the BLS methodology to counting workers in alternative work arrangements is that it only asks workers about their "primary" jobs—thus not capturing workers who work in nonstandard ways in a secondary job or to earn supplemental income or who do not consider their nonstandard work a job at all. The Federal Reserve's Enterprising and Informal Work Activities Survey, fielded in late 2015, sheds light on this issue. It determined that more than one-third of

---

[1] See, for example, Susan Caminiti, "4 Gig Economy Trends That Are Radically Transforming the US Job Market," CNBC.com, October 29, 2018; and Jay Shambaugh, Ryan Nunn, and Lauren Bauer, "Independent Workers and the Modern Labor Market," Brookings Institution (website), June 7, 2018.

[2] BLS, "Contingent and Alternative Employment Arrangements—May 2017," press release, USDL-18-0942, June 7, 2018d.

[3] Lawrence F. Katz and Alan B. Krueger, "The Rise and Nature of Alternative Work Arrangements in the United States, 1995–2015," National Bureau of Economic Research Working Paper No. 22667, September 2016. The difference between the BLS and Katz and Krueger findings may reflect changes in the economy between 2015 and 2017 that resulted in fewer workers in alternative work arrangements as the national economy has strengthened. It also could be due, in part, to statistical variation and the smaller sample size for the RAND ALP.

Americans age 18 years and over had engaged in "informal online and offline paid work activities," and 56 percent of these individuals considered themselves to be formally employed (i.e., employed in a job in addition to performing informal work for pay).[4]

Appendix C includes a further discussion of how the academic research categorizes nonstandard work arrangements and a review of what the literature says about the potential benefits and drawbacks of nonstandard work arrangements in various settings.

Examples of Nonstandard Work Models in the Private Sector

Drawing on the various categorization schemes discussed in this chapter, we developed a typology of flexibility categories, which can be divided into subtypes. The major categories of work arrangements in our typology were as follows:

- flexible scheduling arrangements in the context of an essentially full-time workforce, such as flextime or a compressed workweek
- flexible scheduling arrangements in the context of a part-time or variable workforce, such as seasonal, fixed-block, evening or weekend, shared, or on-call work
- flexible location arrangements, such as telework, remote work, or roaming among sites
- leave flexibility, such as paid leave under the Family and Medical Leave Act (FMLA), unpaid leave, or sabbaticals
- benefits and compensation flexibility, such as childcare or incentive compensation

---

[4]   Barbara Robles and Marysol McGee, "Exploring Online and Offline Informal Work: Findings from the Enterprising an Informal Work Activities (EIWA) Survey," Finance and Economics Discussion Series 2016-089, Washington, D.C.: Board of Governors of the Federal Reserve System, 2016.

- flexible terms of service or duration of relationship, such as recurring variable work, work on retainer, contract work, or temporary employment
- other novel arrangements, such as platform technologies.

In this section, we present examples of private-sector models that fit within this typology and that in some cases represent subtypes of flexibilities within the broader categories. It should be noted that models can include multiple types of flexibilities, for example, allowing both flexible work scheduling and flexible work location.

Flexible Schedule: Full-Time Workforce

Many private employers offer flexible scheduling arrangements to their full-time workforce, such as flextime programs and compressed workweeks. According to the 2016 National Study of Employers, a nationally representative survey of employers including small and large businesses and for-profit and nonprofit employers,[5]

- 81 percent of organizations allow at least some employees to periodically change their work hours within a range of hours, and 32 percent allow all or most employees to do so
- 42 percent allow at least some employees to change their hours on a daily basis, and 11 percent allow all or most of their employees to do so
- 43 percent offer compressed workweeks to at least some employees, and 9 percent offer these arrangements to all or most employees.

The American Time Use Survey considers similar questions from the perspective of the employee. According to that survey, about half of workers report having flexibility in scheduling their hours, and 40

---

[5] Kenneth Matos, Ellen Galinsky, and James T. Bond, *2016 National Study of Employers*, Society for Human Resources Management, Families and Work Institute, and When Work Works, 2017.

percent report having flexibility in the days they work.[6] The share of workers with flexibility to vary their hours differs across industries, for example, with about one-third of workers in the construction and transportation and utilities having flexibility to vary their hours and about two-thirds of information services industry workers having this flexibility.

Another form of a flexible schedule is a results-only work environment (ROWE), which also overlaps with flexible locations, discussed below. In such an arrangement, "[j]ob performance is evaluated solely on the basis of whether the necessary results are achieved by employees, not whether they've put in 'face-time' at the office."[7] Ryan LLC, a tax services firm, is an example of one firm that has implemented a ROWE.[8] However, one of the early adopters of ROWE, Best Buy, has since ended the program for its employees.[9]

Some employers utilize innovative tools to facilitate flexible scheduling practices, such as scheduling analytics software, mobile scheduling technologies, and allowing self-scheduling and shift swaps.[10] In the health care industry, for example, there are hospitals that use software and predictive analytics to forecast patient volume and have allowed workers to alter their schedules to maintain sufficient support at peak times while allowing workers more flexibility to build their schedules

---

[6]   Council of Economic Advisers, *Work-Life Balance and the Economics of Workplace Flexibility*, Washington, D.C.: Executive Office of the President, June 2014b.

[7]   Jason Colquitt, Jeffrey Lepine, and Michael Wesson, *Organizational Behavior: Improving Performance and Commitment in the Workplace*, 4th ed., New York: McGraw-Hill Education, 2015, cited in Steve Nguyen, "Results-Only Work Environment (ROWE)," *Workplace Psychology*, January 4, 2017.

[8]   Lauren Dixon, "How Ryan Flipped to Flex," *Talent Economy*, December 14, 2016.

[9]   Nguyen, 2017.

[10]   Jaime Leick and Kenneth Matos, *Workflex and Health Care Guide*, When Work Works, March 2017; Jaime Leick and Kenneth Matos, *Workflex in Retail, Service, and Hospitality Guide: Cooperative Scheduling, Beyond Bias*, When Work Works, 2016; Liz Watson and Jennifer E. Swanberg, *Flexible Workplace Solutions for Low-Wage Hourly Workers: A Framework for a National Conversation*, Workplace Flexibility 2010 at Georgetown Law School and Institute for Workplace Innovation at the University of Kentucky, May 2011.

outside of core hours.[11] Retailers are experimenting with scheduling applications that allow workers to swap shifts and pick up additional shifts.[12] Even some manufacturers, typically thought to have more strict shift schedules because of the sequential nature of the work, are reevaluating shift rotation schedules and implementing compressed workweeks and other flexible arrangements, typically on a team rather than individual basis.[13]

Flexible Schedule: Part-Time or Variable Workforce

Scheduling practices can vary even more substantially when it comes to part-time or variable workforces, such as seasonal workers, workers in job-sharing arrangements, and other on-call or intermittent employees.

Seasonal work is common in a number of industries. The BLS has identified the following industries as having summer employment peaks: hotels and motels, fitness and recreational sports centers, gold courses and country clubs, amusement parks and arcades, museums, historical sites and zoos, marinas, and recreational and vacation camps.[14] Industries with winter employment peaks include temporary help services, many categories of retail stores, electronic shopping, tax preparation services, and skiing facilities.[15]

One example of an employer that hires a surge workforce to meet seasonal demand is H&R Block, which hires tens of thousands of workers to supplement their core staff at headquarters and retail establishments across the country during tax season.[16] These positions require up-front education and training, including passing the H&R Block Tax Knowledge Assessment, and ongoing education and train-

---

[11] Leick and Matos, 2017.

[12] Watson and Swanberg, 2011.

[13] Kenneth Matos and Eve Tahmincioglu, *Workflex and Manufacturing Guide: More Than a Dream*, When Work Works, 2015.

[14] BLS, "Industries with Summer Employment Peaks," *The Economics Daily*, July 15, 2014a.

[15] BLS, 2014a.

[16] H&R Block, *2017 Annual Report*, Kansas City, Mo., 2017b.

ing for those who return year after year.[17] H&R Block seeks seasonal employees with flexibility to work evenings and weekends, and many seasonal tax preparers also have full-time, year-round jobs. Pay is based on commission (number and complexity of returns) in addition to a modest base wage.[18] Other major examples of employers hiring seasonal surge workers include parcel delivery companies and online and brick-and-mortar retailers that hire additional workers around the holidays, including UPS, FedEx, Target, and Amazon.[19]

A related model is PwC's Flexibility^2 Talent Network, "a group of experienced individuals available during peak periods" that supplements the company's traditional employees.[20] According to PwC materials, this program was developed in part to attract millennials and others who seek greater flexibility in their schedules and careers, and participants can include individuals who "might have a seasonal beach side business, are passionate about volunteering or need to care for a family member."[21] The positions typically last no more than four to six months, can be either full time or part time during that period, and are "paid competitively based on market conditions," potentially including completion bonuses.[22] This program is just one of a wide range of flexible work arrangements offered by PwC,[23] which conducted a study

---

[17] H&R Block, "A Little About H&R Block," fact sheet, 2017a.

[18] Dona DeZube, "Seasonal Tax-Preparation Jobs," *Career Advice* (blog), Monster.com, undated.

[19] Charisse Jones, "UPS and FedEx Will Hire Thousands This Holiday Season," *USA Today*, September 20, 2017; Kellie Ell, "Want a Holiday Job? Here's Who's Hiring Seasonal Employees," *USA Today*, October 25, 2017; Paul Turner, "My Two Months of Seasonal Work at an Amazon Fulfillment Center," *Billfold*, February 4, 2016.

[20] PwC, "Flexibility^2 Talent Network," fact sheet, 2015.

[21] PwC, 2015; PwC, "PwC's Flexibility^2 Talent Network—The Best of Both Worlds," webpage, undated(a).

[22] PwC, "PwC's Flexibility^2 Talent Network FAQs," webpage undated(b).

[23] PwC, "Quality of Life: Balancing Work-Life Demands," webpage, undated(c).

of how it could better recruit and retain workers in response to self-described "crisis-level attrition."[24]

More generally, across occupations, about 15 percent of workers in 2017 usually worked a part-time schedule (excluding those who work part time but who would prefer full-time work), including about one-quarter of service occupation workers and one in five in sales and office occupations.[25] Some employers use "just-in-time" scheduling practices to match part-time labor with workflow demands, giving employees little input and little notice regarding their schedules.[26] Early-career workers may be particularly prone to schedule unpredictability.[27] However, there are also employers that have developed models to incorporate part-time workers in flexible ways while mitigating the instability that is often associated with part-time work. For example, schedule planning and swap tools used to manage full-time workforces can also be adopted for part-time workforces. Employers may give workers greater say in which days they are available to be scheduled, solicit volunteers and offer incentive pay for hard-to-fill or overtime shifts, or offer core workdays to part-time workers with only one day that they may have to be on call. Cross-training, wherein employees are trained across a range of tasks, can help to facilitate scheduling flexibility, as can utilizing float or relief pools to cover gaps.

Job sharing, which involves one or more individuals sharing one full-time job, is a unique form of part-time or intermittent work

---

[24] Dennis Finn and Anne Donovan, *PwC's NextGen: A Global Generational Study, 2013, Summary and Compendium of Findings*, PwC, 2013; Corrine Purtill, "PWC's Millennial Employees Led a Rebellion—and Their Demands Are Being Met," *Quartz at Work*, March 20, 2018.

[25] BLS, "Persons at Work by Occupation, Sex, and Usual Full- or Part-Time Status," *Labor Force Statistics from the Current Population Survey*, last modified January 19, 2018.

[26] Lonnie Golden, "Irregular Work Scheduling and Its Consequences," Washington, D.C.: Economic Policy Institute, Briefing Paper 394, April 2015.

[27] Susan J. Lambert, Peter J. Fugiel, and Julia R. Henly, *Precarious Work Schedules Among Early-Career Employees in the US: A National Snapshot*, Chicago, Ill.: Employment Instability, Family Well-Being, and Social Policy Network, University of Chicago, August 27, 2014.

arrangement. There are only limited data available on job sharing.[28] The 2016 National Study of Employers found that 19 percent of employers offered job sharing to at least some employees while just 2 percent offered these arrangements to most or all employees.[29] Older workers, caregivers, and workers with work-life conflicts may take advantage of job-sharing arrangements to stay connected to the labor force despite being unable to manage the demands of a full-time job. The Society for Human Resource Management (SHRM), a professional organization that studies HR issues, has identified a number of individual businesses, large and small, that offer job sharing among their flexible workplace arrangements.[30]

Flexible Location

Alternative work arrangements may also involve flexibility on where the work is performed. The spectrum of flexible location arrangements ranges from occasional telework during "emergencies or other specific employer-approved situations" to regular telework to full-time telework.[31] There can also be a range of allowable locations for the work, with employees having full discretion over where they work (at home or another location) or with off-site work occurring at a central location, while in transit, or at a client's location.

The 2016 National Study of Employers found that 66 percent of employers allow at least some employees to work at home occasionally, and 40 percent allow at least some employees to work at home on a regular basis.[32] The reported shares of employers allowing all or most employees to work at home are considerably lower, with 8 percent allowing all or most employees to work at home occasionally and 2

[28] Elka Maria Torpey, "Flexible Work: Adjusting the Where and When of Your Job," *Occupational Outlook Quarterly*, Summer 2017.

[29] Matos, Galinsky, and Bond, 2017.

[30] When Work Works, "Workplace Awards," webpage, undated.

[31] Kenneth Matos, *Workflex and Telework Guide: Everyone's Guide to Working Anywhere*, When Work Works, 2015.

[32] Matos, Galinsky, and Bond, 2017.

percent allowing all or most employees to work at home on a regular basis. The BLS American Time Use Survey, which surveys individuals rather than employers, offers another perspective on working from home. According to 2017 annual averages, about 20 percent of wage and salary workers and 55 percent of self-employed workers worked from home for at least some amount of time on an average day.[33] The share working from home varies by occupation, with more than 30 percent of workers in management, business, and financial occupations and professional and related occupations working from home on an average day; by contrast, less than 10 percent of workers in these occupations worked from home on an average day: service; construction and extraction; installation, maintenance, and repair; production; and transportation and material moving.

According to SHRM, a major impetus for utilizing flexible location work arrangements is to broaden the pool of human capital that the employer can access: "Telework can open entirely new hiring markets to your recruiters by removing or reducing the obstacles placed by distance, mobility-based disabilities and some conflicting family responsibilities."[34] A recent survey of hiring managers found that those at organizations with remote workers are more likely to think that hiring has become easier, and on balance these hiring managers prioritize skills over whether or not a prospective hire would work at the same location as the rest of the team.[35]

However, some tasks are more suitable to being performed off-site than others, as suggested by the differential shares of workers by occupation who report working from home. SHRM describes four "essential characteristics" that make work suitable to being done remotely:[36]

1.   Do job tasks require onsite-only resources?

---

[33] BLS, "American Time Use Survey—2017 Results," press release, USDL-18-1058, June 28, 2018e, Table 7.

[34] Matos, 2015.

[35] Inavero, "2018 Future Workforce Report: Hiring Manager Insights on Flexible and Remote Work Trends," presentation slides, February 2018.

[36] Matos, 2015.

2. Is walk-in customer service a primary responsibility?
3. Does the employee have sufficient independent access to information?
4. Does the position allow sufficient autonomy for the employee to work remotely?

OPM materials note that "most jobs, if not all, include some duties that are considered 'portable' in that they generally can be performed from any location."[37] In particular, "writing, analysis, and research tasks are ideally suited to the quieter, less distracting environment of the alternate site."[38] Both OPM and SHRM make the case for delineating between tasks suitable to telework and tasks that are not and seeking to provide employees with flexibility to telework when appropriate.

Technology is a major facilitator of remote work arrangements, allowing for collaboration in real time across locations. The General Services Administration (GSA), for example, lists tools that can aid telework, including Thin Client, virtual private networks that allow employees to "tunnel" in to the employer's network from a remote computer, online meeting and messaging tools, and voice over IP technologies to permit phone calls through a computer.[39] Technology advances also offer the potential to expand the set of tasks that are suitable to telework. For example, high-tech, high-resolution cameras are contributing to the rise in telehealth and telemedicine.[40] Job-search websites show that a number of health care companies are hiring for physicians and other medical professionals to work remotely.[41] Some companies operate strictly on a remote consultation basis, for example, RediDoc,

---

[37] Telework.gov, "Telework Managers: Assessing Job Tasks," webpage, OPM, undated(b).

[38] Telework.gov, undated(b).

[39] U.S. General Services Administration, "Tools for Effective Telework," webpage, undated.

[40] Leick and Matos, 2017; E. Ray Dorsey and Eric J. Topol, "State of Telehealth," *New England Journal of Medicine*, Vol. 375, No. 2, July 14, 2016.

[41] See, for example, Flexjobs.com, "Doctor and Practitioner Remote, Part-Time, and Freelance Jobs," webpage, undated(a); Indeed.com, search for "Telemedicine Physician Work from Home," undated; Laureen Miles Brunelli, "Medical Jobs from Home," *The Balance Careers*, June 9, 2017.

which "securely connects patients with U.S. based, board certified physicians for medical consultations via telephone, secure video, and secure email."[42] Such "direct-to-consumer" telehealth services "appear to be accelerating" according to recent research.[43]

Flexible location arrangements can also involve working at locations within the company that vary over time, seasonally or based on demand. CVS Caremark, for example, had offered a "snowbird" program that allowed employees to transfer from retail establishments in colder regions to establishments in warmer regions on a seasonal basis, as part of an effort to retain older workers and adjust to seasonal fluctuation in demand across locations (an initiative that has since been incorporated into broader flexible work arrangement policies).[44] Another approach is to allow workers to work at different establishments in the same general area, with the flexibility to move across locations helping employees to maintain more consistent hours than could be offered at just one location.[45]

Leave Flexibility

Employers may also provide employees with flexibility to alter or reduce their schedules over the course of their careers, for example, by providing paid or unpaid family and medical leave, by allowing employees to ease back into full-time schedules following childbirth, or by offering employees the flexibility to take sabbaticals and return to their position.

The share of workers with access to paid leave policies varies across data sources. According to BLS's 2016 National Compensation Survey,

---

[42] RediDoc, homepage, undated.

[43] J. Scott Ashwood, Ateev Mehrotra, David Cowling, and Lori Uscher-Pines, "Direct-to-Consumer Telehealth May Increase Access to Care but Does Not Decrease Spending," *Health Affairs*, Vol. 36, No. 3, 2017.

[44] Sloan Center on Aging and Work, "CVS Caremark Snowbird Program," Innovative Practices Database, 2012; Steven Greenhouse, "The Age Premium: Retaining Older Workers," *New York Times*, May 14, 2014; Rick Wartzman, "What America's Aging Workers Mean for the Future of Work," *Fortune*, June 22, 2016.

[45] Watson and Swanberg, 2011.

14 percent of private-sector workers have access to paid parental or family leave, with workers in traditionally white-collar fields having much greater access to paid leave than workers in service industries.[46] By contrast, the American Time Use Survey Leave Module from 2011 found that about 40 percent of workers reported having access to some type of paid leave after childbirth.[47] The 2016 National Survey of Employers found that 58 percent of employers report providing at least some paid maternity leave, and 15 percent report providing at least some paid paternity leave.[48] Several of these sources found higher shares of paid leave in larger businesses than at smaller firms.[49] According to a report by the Boston Consulting Group, higher-paid workers with specialized skills are more likely to have access to paid leave in their jobs, as employers use it as a benefit to attract and retain talent.[50]

Unpaid family and medical leave is both more common than paid leave and varies less across occupations and wage levels, due in large part to 1993's FMLA.[51] Other forms of leave include sabbaticals or career breaks. According to the National Survey of Employers, 28 percent of employers allow at least some employees to take sabbaticals of six months or more and return to work, and 11 percent allow all or most employees to do so, while 55 percent of employers allow at least some employees to take "extended career breaks for caregiving or other personal or family responsibilities," and 35 percent of employers allow all or most employees to take such breaks.[52] According to this same survey, 81 percent of employers allow at least some employees to

---

[46] Drew Desilver, "Access to Paid Family Leave Varies Widely Across Employers, Industries," *Fact Tank* (blog), Pew Research Center, March 23, 2017.

[47] AEI-Brookings Working Group on Paid Family Leave, *Paid Family and Medical Leave: An Issue Whose Time Has Come*, Washington, D.C., May 2017.

[48] Matos, Galinsky, and Bond, 2017.

[49] Desilver, 2017; Matos, Galinsky, and Bond, 2017.

[50] AEI-Brookings Working Group on Paid Family Leave, 2017.

[51] Council of Economic Advisers, *The Economics of Paid and Unpaid Leave*, Washington, D.C.: Executive Office of the President, June 2014a.

[52] Matos, Galinsky, and Bond, 2017.

return to work gradually after childbirth or adoption, and 52 percent of employers allow all or most employees this flexibility. One innovative model regarding easing back into work following childbirth is Amazon's Ramp Back program, which offers employees the option to work at half or three-quarter time for up to eight weeks.[53]

Innovative models under the umbrella of leave flexibilities also exist when it comes to allowing employees to customize their career paths, working more or less from year to year as family, educational, and other demands arise. One example is Deloitte's "career lattice" or "mass career customization" model, which is designed as a shift away from a typical "career ladder" and allows for multiple pathways to move horizontally or diagonally within the organization, adjusting work pace, workload, location/schedule, and role over time.[54] Other employers have embraced the shift from "ladder" to "lattice" as well.[55]

U.S. Non–Department of Defense Public-Sector Organizations

Many nonstandard work arrangements in use in the private sector are utilized by public-sector organizations as well. Below, we describe alternative work arrangements in use across nondefense federal agencies and in state and local governments.

Other Federal Agency Models

The OPM handbook *Human Resources Flexibilities and Authorities in the Federal Government* outlines hiring authorities that federal agencies

---

[53] Nitish Kulkarni, "Amazon Revamps Parental Leave Policy," Techcrunch.com, November 2, 2015; Steve Winter, "Two Years in, Our Parental Leave Policy Is Working for Parents," *dayone: The Amazon Blog*, August 21, 2017.

[54] Ann Weisberg, Deloitte, "Mass Career Customization: Building the Corporate Lattice Organization," presentation at Work/Life Policy, Practice and Potential, UN Expert Group Meeting, November 9, 2010.

[55] Peter Cappelli and J. R. Keller, "Talent Management: Conceptual Approaches and Practical Challenges," *Annual Review of Organizational Psychology and Organizational Behavior*, No. 1, March 2014.

can use to draw in workers on a short-term basis or through alternative staffing options.[56] Short-term hiring authorities include the following:

- Intergovernmental Personnel Act appointments for a maximum of two 2-year terms of service (5 U.S.C. 3371–3375; 5 CFR [Code of Federal Regulations] part 334)
- temporary, competitive appointments not to exceed 1 year in duration (5 CFR part 316, subpart D)
- term, competitive appointments of 1 to 4 years, where the need for the employee is not permanent but driven by short-term work-load demands (5 CFR part 316, subpart D)
- intra-agency details of up to 120 days (5 U.S.C. 3341)
- employing experts or consultants on a temporary or intermittent basis, with appointment terms not to exceed 1 year (5 U.S.C. 3341)
- commercial temporary help services for a maximum of two 120-day periods (5 U.S.C. 3341)
- contracting, following federal procurement regulations.

Alternative staffing options include the following authorities:

- programs for students and recent graduates, including the Internship Program, Recent Graduates Program, and Presidential Management Fellows Program (5 U.S.C. 3341)
- direct hire authorities, when public notice has been given and OPM determines that there is "a severe shortage of candidates or a critical hiring need" (5 U.S.C. 3304 and 5 CFR part 337, subpart B)
- special appointing authorities: Excepted Service, Schedule A and Schedule B, government-wide or agency specific (5 CFR part 213)
- job sharing and other permanent positions that are not full time, such as part-time, intermittent, and seasonal work (5 CFR 340)

---

[56] OPM, *Human Resources Flexibilities and Authorities in the Federal Government*, Washington, D.C., August 2013.

- reemploying annuitants without salary offset (5 U.S.C. 8344(i) and 8468(f); 5 CFR part 553).

There are additional flexibilities offered for hiring cyber professionals, including at least 28 staffing options that are available to DoD.[57] The federal government also allows for a range of compensation flexibilities to attract and retain cyber professionals.[58]

In addition to hiring flexibilities, OPM details work arrangements and work-life policies that can introduce flexibilities into federal jobs. Scheduling flexibilities include the ability for agencies to establish full-time, part-time, intermittent, and seasonal work schedules; traditional or rotating shift schedules; paid and unpaid breaks during the workday; and alternative work schedules, such as flexible or compressed work schedules. These alternative work schedule arrangements are described in detail in separate OPM guidelines, the *Handbook on Alternative Work Schedules*.[59] A flexible work schedule is defined as one that "allows an employee to determine his or her own schedule within the limits set by the agency," and a compressed work schedule is defined as one that "in the case of a full-time employee, an 80-hour biweekly basic work requirement that is scheduled by an agency for less than 10 workdays."[60]

A set of models of types of flexible or compressed work schedule arrangements is detailed in the OPM materials, each of which have their own terms and conditions. For flexible work schedules, they include Flexitour, gliding schedules, variable-day schedules, variable-week schedules, and MaxiFlex. For compressed work schedules, they include a four-day work week, a three-day work week, and a 5/4/9 compressed plan (eight nine-hour days and one eight-hour day per pay period). The authority for flexible work arrangement programs was

---

[57] DoD, Defense Civilian Personnel Advisory Service, "Cyber Hiring Options/Authorities Guide," December 2016.

[58] OPM, *Compensation Flexibilities to Recruit and Retain Cybersecurity Professionals*, Washington, D.C., undated(a).

[59] OPM, *Handbook on Alternative Work Schedules*, Washington, D.C., undated(b).

[60] OPM, undated(b).

established by the Federal Employees Flexible and Compressed Work Schedules Act and is detailed in code at 5 U.S.C. 61.[61]

In addition to offering flexible work schedules, the federal government offers programs that allow flexible work locations, such as telework. These policies date back at least to 1990, when Congress authorized the Federal Flexplace Project, which allowed agencies "to use appropriated funds to install telephone lines, necessary equipment, and to pay the monthly charges in a private residence."[62] The current regulations governing telework across the federal government have their origin in the Telework Enhancement Act of 2010 (P.L. 111-292), which required all agencies to establish policies under which eligible employees may be authorized to telework and to determine which employees are eligible.[63] This legislation also required annual OPM reports to Congress on the status of telework arrangements across the federal government. The most recent report, covering FY 2016, found that 42 percent of federal employees were eligible to telework in FY 2016 and that about half of these employees did so.[64] While OPM reports that the share of employees eligible to telework has changed little in recent years, the share of eligible employees who do in fact telework has increased consistently, from 29 percent in FY 2012 to 51 percent in FY 2016.[65]

A recent set of reports investigated telework practices in use across seven federal agencies that handle classified or sensitive information and developed a set of recommendations for intelligence community organizations seeking to expand their telework programs in part to improve recruitment and retention of millennial employees who "demand more-flexible work hours, greater ability to work from any

---

[61] Georgetown Federal Legislation Clinic, "The Federal Employees Flexible and Compressed Work Schedules Act (FEFCWA)," Workplace Flexibility 2010 at Georgetown University, 2006.

[62] Telework.gov, "Telework Legislation," webpage, OPM, undated (a).

[63] Public Law 111-292, Telework Enhancement Act of 2010, December 9, 2010.

[64] OPM, *Status of Telework in the Federal Government: Report to Congress Fiscal Year 2016*, Washington, D.C., November 2017.

[65] OPM, 2017.

location, and access to technology not currently allowed in SCIFs."[66] The researchers concluded that "a clear understanding of the purpose of such a program is essential for leaders who will establish the program goals, policies around different parameters, and performance measures, as well as for the managers who will be responsible for developing and implementing new technology capabilities, security protocols, and training."

Other types of workplace flexibilities detailed in OPM materials include but are not limited to paid leave programs, child- and elder-care assistance, subsidized transportation, and a range of flexible compensation practices including recruitment, retention, and relocation incentives.[67]

Of course, flexibility in federal employment can also cut the other way—making jobs more precarious rather than more desirable. For example, the President's Management Agenda of 2001 encouraged agencies to use flexible workers as a way to cut costs—contributing to the "core" and "ring" model of federal employment described in Appendix C.[68]

As of March 2018, of the nearly 2.1 million federal civilian employees in executive branch agencies (excluding the U.S. Postal Service), more than 100,000 people worked on a less than full-time basis, and another more than 35,000 worked full time but only seasonally.[69] About 132,000 people were in nonpermanent positions, with nearly 50,000 of these workers overlapping with the less than full-time schedule category and another nearly 5,000 overlapping with

---

[66] See Cortney Weinbaum, Bonnie L. Triezenberg, Erika Meza, and David Luckey, *Understanding Government Telework: An Examination of Research Literature and Practices from Government Agencies*, Santa Monica, Calif.: RAND Corporation, RR-2023-OSD, 2018; Cortney Weinbaum, Arthur Chan, Karlyn D. Stanley, and Abby Schendt, *Moving to the Unclassified: How the Intelligence Community Can Work from Unclassified Facilities*, Santa Monica, Calif.: RAND Corporation, RR-2024-OSD, 2018. The first of the two reports cited here includes a review of the scholarly literature on the costs and benefits of telework arrangements.

[67] OPM, 2013.

[68] Sharon H. Mastracci and James R. Thompson, "Who Are the Contingent Workers in Federal Government," *American Review of Public Administration*, Vol. 39, No. 4, July 2009.

[69] OPM, "Employment Cubes," FedScope database, last updated March 2018.

the full-time seasonal schedule category. Overall, across the agencies, more than 200,000 people were either part-time employees, temporary employees, seasonal employees, or fell under some combination of these arrangements.

Two federal agencies that have used flexible staffing arrangements to supplement their full-time workforces are NIFC/USFS[70] and the Small Business Administration's (SBA) Office of Disaster Assistance (ODA). NIFC/USFS hires two classes of intermittent surge personnel, "casuals" and seasonal employees. Casuals tend to be retired civil servants with firefighting qualifications, who are hired under an administratively determined pay plan mechanism and are paid only when activated. Seasonal employees are often younger people looking to gain experience and develop their skills, and they are hired noncompetitively under authorities that permit them to work up to 1,040 hours in a year. ODA employs intermittent workers in two key capacities: as *surge* personnel (term-intermittent personnel hired on a competitive basis who are placed on a roster and expect to work on an on-call basis) and *surge plus* personnel (excepted personnel hired as needed only when other personnel categories are insufficient to meet demand). A detailed discussion of how these agencies recruit, retain, and manage these variable workforces is included in Appendix C.

State and Local Government Models

Other levels of government also utilize nonstandard work arrangements to meet their needs. Indeed, past research using CWS data shows that nonstandard workers make up a larger share of state and local workforces than they do the federal workforce.[71] This finding

---

[70] NIFC is a consortium of federal agencies with firefighting responsibilities and capabilities. USFS, National Park Service, Bureau of Land Management, U.S. Fish and Wildlife Service, Bureau of Indian Affairs, U.S. Fire Administration, National Oceanic and Atmospheric Administration, and the National Association of State Foresters participate in NIFC. USFS is closely associated with resourcing and management of NIFC; accordingly, we treat USFS and NIFC as essentially the same entity.

[71] Sharon H. Mastracci and James R. Thompson, "Nonstandard Work Arrangements in the Public Sector: Trends and Issues," *Review of Public Personnel Administration*, Vol. 25, No. 4, December 2005.

is driven principally by the substantial use of part-time, on-call, and seasonal workers in education—for example, teachers, teachers' aides, school administrators, and support staff (substitute teachers count as on-call workers). Police and detectives, correctional institution officers, and construction trade workers also contributed to the high shares of nonstandard arrangement workers in the public sector beyond the federal government. State and local governments have also been active in using programs to rehire retirees to fill critical vacancies.[72]

A review of job opportunities on public job search websites reveals that many of the nation's 80,000-plus state and local governments advertise positions as offering flexibilities, such as alternative work schedules, part-time work, and telework.[73] For example, from 2016 to 2017, state and local government jobs were among the five categories of jobs on the site Flexjobs that experienced at least 20 percent growth in remote job listings.[74] According to a 2018 study by the Center for State and Local Government Excellence, which surveyed HR professionals who are members of either the International Public Management Association for Human Resources (IMPA-HR) or the National Association of State Personnel Executives, about half of state and local governments offer flexible scheduling practices, about 20 percent offer regular telework for certain positions, and 8 percent offer job sharing.[75] Some officials who responded to the survey indicated that their organizations have used gig economy arrangements to fill short-term needs in a range of occupational areas, including IT support, office support, accounting, maintenance work, and consulting services. One in three of the HR professionals surveyed indicated that "creating a more flexible workplace (e.g., job sharing, outsourcing, hiring retirees)" is among the workforce issues important to their organization.

---

[72] Stuart Greenfield, "Public Sector Employment: The Current Situation," Center for State and Local Government Excellence, undated.

[73] Flexjobs.com, undated(a).

[74] Brie Weiler Reynolds, "5 of the Fastest-Growing Remote Career Categories," Flexjobs. com, January 3, 2018.

[75] Center for State and Local Government Excellence, *Survey Findings: State and Local Government Workforce: 2018 Data and 10 Year Trends*, Washington, D.C., May 2018.

State and local government agencies are also among the employers recognized in a database of workplace flexibility programs compiled by SHRM.[76] Types of programs in use across the country include flex place with laptops and an ability to forward work email to their smartphones (Eagle County Colorado Government), full-time work from home positions and the ability to switch from full-time to part-time hours (Arlington County Virginia Human Resources Department), a policy that allows employees to bring their infants to work for the first six months (Arizona Health Care Cost Containment System), compressed work schedules with the ability to vary starting and stopping times (Michigan Occupational Safety and Health Administration), and reducing work hours as employees approach retirement (Rhode Island Housing), among other programs.

Interviews with subject-matter experts in personnel management in state and local governments provided further evidence that these organizations are utilizing flexible work arrangements to improve recruitment and retention. Examples ranged from expanding telework, to allowing scheduling flexibilities to promote employee wellness and work-life balance, to using administrative flexibilities to hire workers with hard-to-recruit knowledge and skills (such as IT or financial management skills) for term-limited positions outside of the standard hiring system. Several officials we interviewed indicated that an expanding economy coupled with evolving work preferences, in particular among members of the millennial generation, will require public-sector organizations to be increasingly innovative in how they develop and deploy flexible work arrangements to draw talent into their workforces. One official indicated that older workers are also seeking pathways to ramp down their workload, suggesting that developing new models of work as baby boomers reach traditional retirement age could help to keep these workers attached to the labor force and contributing to meet organizations' needs.

While generally expressing positive views of how nonstandard work arrangements can help to draw in talent, interviewees noted that

---

[76] When Work Works, undated.

these arrangements can pose challenges as well, such as perceptions of unfairness if some workers are able to take advantage of scheduling and location flexibilities but not others, and difficulties maintaining connections with colleagues. Rules and regulations, including labor laws and agreements with public-sector labor unions, can also limit flexibilities that state and local governments can utilize. So, too, can budgeting procedures, with one interviewee citing head-count caps established in the budget as an impediment to adopting a job-sharing model.

Application to Alternative Reserve Component Management Approaches

This section has described a wide range of nonstandard work arrangements in use in the private and public sectors, including models that involve flexibilities in terms of when work is performed, where work is performed, and the administrative relationship between the organization and the individual performing the work (e.g., whether the individual is an employee or an independent contractor). Although definitive data on the share of workers in nonstandard arrangements are hard to come by due to definitional and methodological challenges, there is a widespread perception that these arrangements are becoming more prevalent, the result of both workers and employers seeking greater flexibility.

Some of the flexible workforce models are already being utilized at least to some extent by the RCs (e.g., several have adopted policies that permit telework to a limited degree),[77] while others may face insurmountable legal or bureaucratic obstacles to adoption (e.g., there are limits on the type of work contractors can perform under current laws and policies). However, many of the models, or permutations of them, could help to break down time and location obstacles to reserve service (the two axes of the conceptual model presented in Chapter Four), allowing the RCs to draw in people who may otherwise be unable to

---

[77] T. Scott Randall, "Can I Drill from Home? Telework (or the Lack Thereof) in the Army Reserve," *Army Lawyer*, April 2014.

participate or to use current members in new ways that expand beyond the traditional service model.

In the next chapter, we connect the themes presented thus far in this report, drawing on knowledge of existing shortages in the RCs, populations that are underutilized under current service constructs, and innovative models to employ human capital in use in foreign militaries and the public and private sectors to outline a set of alternative constructs.

CHAPTER SEVEN

Potential Workforce Constructs for Innovative Reserve Component Workforce Management Models

The previous chapters have been focused, respectively, on three components of the problem at hand: workforce demand, workforce supply, and potential alternative workforce models. In this chapter, we try to overlay all three and identify potential workforce constructs for innovative human resource management, at the intersection of demand, supply, and alternative models.

The potential workforce constructs that follow are not meant to be definitive. They were developed in the iterative process described in Chapter One, which ended with the research team devising several new, RC-focused program concepts to address the intersections of demand, supply, and alternative models. If a research team member identified, for example, an underrepresented population in the current RC workforce (supply), they were asked to think about what military functions in demand they might be most profitably managed to perform. Through these multiple iterations, the research team reviewed the list of alternative constructs to ensure that, as a group, they addressed as many critical specialty demands as possible, targeted the most clearly underrepresented populations, and leveraged the most promising models from other countries and other sectors. It was not within the scope of this research to devise a workforce construct for each combination of workforce demand, workforce supply, and alternative workforce models, but the key findings on each of these three dimensions are found in at least one construct.

The alternative constructs presented in this chapter are also not mutually exclusive. For example, interviewees whom we spoke with about these constructs in relation to appealing to single parents stated that the constructs for Telereserves and Reserves on Demand could be particularly appealing in combination.[1] Finally, while we outline potential disadvantages with each construct, such discussions were primarily informed by limitations that were easily identifiable in our research. It is critical to note that each construct might entail additional risks and disadvantages beyond those discussed below.

Table 7.1 shows how the various constructs have primary effects on different military requirements and target populations. The challenge of time is divided into two subsets, as some civilian employment varies by the hour or day within a broader general pattern (a firefighter works the same number of hours every month, but those hours may fall on different days each month), and in other cases it varies from month to month (either seasonally or with other cycles, like oil rig crews). RAND did not identify a model for alternative manpower sources that would substantially address shortages in two areas: special operations forces and EOD/CBRN. Due to their highly specialized and military-centric natures, these will likely remain hard to fill through either conventional or alternative workforce constructs.

Workforce Construct 1: No Passport Required

What Is This Workforce Construct?

This construct seeks to expand the population of eligible potential service members by eliminating for selected individuals the requirement to be available for worldwide deployment.

An increasing number of potential recruits are unable to meet the military's minimum accession physical or medical standards.[2] Family

---

[1]   Telephone interview with representatives from an organization that focuses on issues related to women in the workforce, July 25, 2018.

[2]   Office of the Under Secretary of Defense for Personnel and Readiness, *Population Representation in the Military Services: Fiscal Year 2014 Summary Report*, Washington, D.C.,

Table 7.1
Workforce Constructs at the Intersection of Supply and Demand Targets

Shortage Area	Time (hours and days)	Time (weeks and months)	Distance	Deployability
Cyber, computer/ network technicians	Reserves on Demand		Telereserves	No Passport Required
Intelligence	Reserves on Demand		Telereserves	
Maintenance	Reserves on Demand		Sponsored Reserve	
Aviation	Part-Time Plus	Job Sharing		
Medical professionals		Job Sharing	Telereserves	No Passport Required
Construction		Seasonal Worker, Seasonal Reserve	Sponsored Reserve	
Special operations forces				
EOD/CBRN				
Linguists	Reserves on Demand	Seasonal Worker, Seasonal Reserve	Telereserves	
Chaplains	Warrant Officer-Deacons			
Transportation	Reserves on Demand	Seasonal Worker, Seasonal Reserve	Sponsored Reserve	
General unit support	Part-Time Plus			Wounded Warrior

structures are continuously changing as the number of single mothers and cohabiting unmarried parents increases.[3] Meanwhile, service

March 1, 2016; Joint Advertising Market Research and Studies, *Generational Values and Their Impact on Military Recruiting*, October 2016.

[3] Pew Research Center, *Parenting in America: Outlook, Worries, Aspirations Are Strongly Linked to Financial Situation*, Washington, D.C., December 17, 2015; U.S. Census Bureau, "Mother's Day: May 8, 2016," press release No. CB16-FF.09, April 20, 2016.

members perform a wide range of tasks and require a similarly wide range of physical performance. Some need to kick down doors and be able to carry a large amount of weight on their backs while others need to be able to sit for a shift monitoring a console. Some need to lift and move heavy equipment while others need to control weapons through keyboards and joysticks. However, there is no variation in medical/physical standards even if the typical performance of those specialties are not physically demanding, because of the possibility that any service member may be called upon to deploy, serve in austere conditions, and potentially engage in direct combat. This proposal trades universal deployability to have access to a larger population.

Currently, if someone is unable to meet the basic physical or medical accession standards or needs lifestyle accommodations (single or dual military parents), it is suggested that he or she should consider joining the services as a civilian. Whereas some of these responsibilities and positions could potentially be filled by civilians, engagement in armed conflict has historically been deemed exclusively military based on combatant status under international law.[4] There are also limited part-time defense opportunities for civilians; that is, there is no civilian equivalent of the RCs.

This proposal is a radical departure from the status quo. A February 14, 2018, memo from the Under Secretary of Defense for Personnel and Readiness directed that "[s]ervice members who have been non-deployable for more than 12 consecutive months, for any reason, will be processed for administrative separation," granting an automatic exception only for pregnant or postpartum service members, to increase the readiness and lethality of the force.[5] This proposal would require a change in this policy, at least for a subset of specialties, and a larger change in culture away from the expectation of every service member who has to deploy.

---

[4] See Geneva Convention Relative to the Treatment of Prisoners of War art. 4(A)(1)-(4), Aug. 12, 1949, 6 U.S.T. 3316, 75 U.N.T.S. 135; see also Alex Wagner and Andrew Burt, "Blurred Lines: An Argument for a More Robust Legal Framework Governing the CIA Drone Program," *Yale Journal of International Law Online*, Vol. 38, Fall 2012.

[5] Wilke, 2018.

The RCs are organized, trained, and equipped for deployment and use outside of the United States. However, as described in Chapter Five, countries with similar RCs have organized, trained, and equipped their reserve forces with different assumptions. For example, the French reserve model plans for most reservists to backfill positions in France vacated by other military members deploying outside of the national territory. Although it would be a dramatic change in culture, there may be opportunities to design units that would primarily serve in a similar function as the French model and better leverage some of the populations able to serve in No Passport Required.

Whom Does This Workforce Construct Target?

No Passport Required targets individuals who desire to serve and are available for typical drill periods and annual training requirements, are either unable or unwilling to deploy outside of the United States or in austere environments for extended periods, but are otherwise able to perform the duties of a position. Potential populations may include individuals with well-managed medical conditions that are unsafe in a deployed or austere environment (e.g., severe food allergy, diabetes, sleep apnea, asthma), those unable to meet physical standards (overweight, injured, or disabled), and parents who can arrange short-term childcare but would struggle to arrange longer-term care (single parents or dual-military/dual-career couples).

There appear to be populations that would be interested in serving in this type of model. In an interview with representatives from an organization that focuses on issues related to women in the workforce, we heard that "the biggest issue is predictability" when it comes to those who have childcare needs. One said,

> Nights and weekend aren't that much bigger of a problem because the biggest source of childcare in the U.S. is friends and neighbors or aunts or grandparents or siblings. . . . If you know ahead of time you have to be somewhere every fourth weekend, it's like if you do a kind of distance learning program where you have to

be somewhere every few weekends. So I think that's not necessarily an additional burden.[6]

To Whom Is It Relevant? What Gaps Does This Workforce Construct Aim to Fill?

This construct would be particularly helpful to fill gaps in specialties that are not principally practiced "downrange," such as cyber operations, remotely piloted vehicle crews, and strategic systems, among the obvious examples. These positions could be served with greater access to a larger population that would otherwise be well suited for the position. Many of the specialties best suited for No Passport Required are enabled by developments in technology that allow them to operate at a distance from the area of operation or have substantial stateside requirements in addition to deployment requirements. As technology and the proliferation of weapons engaged in combat from a distance continues to evolve, there is potential for greater parts of the force to be more suitable for this construct.

Potential Disadvantages

Risks of this construct would include (1) the dilution of the current policy focused on minimizing the number of personnel permanently undeployable and monitoring closely the duration of undeployable times for the rest of the force and (2) increased stress on the deployable portion of the force. The former is not a substantial risk, as policies can be and are changed even within one administration, and we are looking for long-term changes. The latter assumes the continuation or resumption of a rotational force posture where even personnel whose specialties are not needed overseas may be called to deploy in branch immaterial positions to reduce the deployment burden on those who have already deployed. To some degree, every person taken off the list of potential deployers has the potential to reduce dwell time for the rest of the force.

---

[6] Telephone interview with a representative from an organization that focuses on issues related to women in the workforce, July 25, 2018.

Workforce Construct 2: Telereserves

What Is This Workforce Construct?

This option would involve an expansion of existing telework arrangements to allow more RC members to take advantage of ongoing advances in technology to perform a broader set of tasks remotely—from their homes, a location closer to home than where they typically would train or perform active-duty service, or at a remote location away from both their home and duty station but where their civilian job requires them to be.

First and foremost, this construct would be designed to break down the location barriers to reserve service discussed in Chapter Four, facilitating participation among individuals who may have the time to serve but for whom the need to be at a certain location to perform that service poses a sufficient obstacle to prevent them from joining or staying in the RC, because of the demands of their civilian job, personal characteristics or responsibilities, or other reasons. A 2017 GAO report, for example, cited several DoD reports and officials that make the case that unreimbursed out-of-pocket travel expenses "may, among other factors, be a challenge for reservists and may therefore negatively affect retention," which suggests that developing models that allow more to be done without traveling long distances could promote retention.[7]

The Telereserves model could also alleviate the scheduling obstacles to reserve service discussed in Chapter Four by cutting down on travel time and by allowing individuals to be in a different location during the portions of the day when they are not training or on duty and therefore to tend to responsibilities that they may have in that location. Combining this model with other workforce constructs discussed in this chapter, for example, the Reserves on Demand or Part-Time Plus models, could enhance its ability to cut through both location and time obstacles to service, enabling individuals to serve at times and places of their choosing. Telereserves also could be used in tandem with the No Passport Required model to target individuals who would

---

[7] GAO, *Reserve Component Travel: DOD Should Assess the Effect of Reservists' Unreimbursed Out-of-Pocket Expenses on Retention*, Washington, D.C., GAO-18-181, October 2017b.

benefit from a service option that both allows them to serve remotely (from home or elsewhere) and does not require them to deploy overseas.

Implementing a Telereserves model by expanding who can telework, how much, and for what purposes, in a broad sense, would be permissible under DoDI 1035.10, which provided overall implementing guidance to DoD related to the Telework Enhancement Act of 2010.[8] However, this DoDI devolved authority for establishing telework policies to the individual RCs, which all RCs with the exception of the Army Reserves have since done, with some RCs establishing more permissive policies than others.[9] Expanding telework will also involve investments in technologies that facilitate it, such as computing and communications technologies, as well as giving careful consideration to how to promote cybersecurity.[10] In the case of telemedicine, such technologies include "digital stethoscopes, camera illumination systems, digital otoscopes, digital blood pressure monitors, etc."[11] One individual we interviewed from an organization that conducts work related to military families stated that telework makes a lot of sense, and it is happening in the rest of the world. However, the interviewee pointed out that from their experience, security would be the biggest challenge. Sending files, having video and audio connections, and having access to defense systems could present practical barriers to the feasibility of this option. This interviewee noted that concerning health care, issues with telework have been figured out elsewhere but not yet in the military world.[12]

In addition to adjusting policies and making investments in necessary resources, a substantial expansion of remote work could involve

---

[8]   DoDI 1035.10, *Telework Policy*, Washington, D.C., April 4, 2012.

[9]   Randall, 2014.

[10] Weinbaum, Chan, et al., 2018; Weinbaum, Triezenberg, et al., 2018.

[11] John E. Whitley, James M. Bishop, Sarah K. Burns, Kristen M. Guerrera, Philip M. Lurie, Brian Q. Rieksts, Bryan W. Roberts, Timothy J. Wojtecki, and Linda Wu, *Medical Total Force Management: Assessing Readiness and Cost, Institute for Defense Analysis*, Washington, D.C.: Institute for Defense Analyses, Paper P-8805, May 2018.

[12] Telephone interview with organization focused on issues related to military families, July 18, 2018.

a reconceptualization of what it means to serve—and a shift away from a model in which units drill together in the same place at the same time to enhance readiness and unit cohesion to one that relies on individuals to perform independently to a greater degree and to collaborate using technology. Certainly, many military tasks will not fit into a remote work model, just as there are occupations in the private sector that are less conducive to remote work, such as service industries. But as the RCs seek to draw in talent from populations that are underutilized in existing workforce constructs, they should keep in mind that location can pose an insurmountable barrier and consider ways to break down that barrier by expanding the range of the possible when it comes to performing service remotely.

Whom Does This Workforce Construct Target?

Critically, the Telereserves workforce construct would be designed to break down location barriers to RC service, facilitating participation among individuals who may have the time to serve but whose civilian jobs or personal responsibilities prevent them from being at a duty location on particular days.

Among the underutilized populations that a Telereserves construct could target are working parents, including single parents who otherwise may be unable to serve. In an interview with representatives from an organization that focuses on issues related to women in the workforce, one interviewee said,

> I think telework could be really helpful for working parents if they are able to do it on their own time schedule. So maybe they are working from—they are able to drop their kids off at day care and start later in the morning, work after the kids have gone to bed, and have the flexibility to get their work done when it makes sense for their caregiving schedules.[13]

---

[13] Telephone interview with a representative from an organization that focuses on issues related to women in the workforce, July 25, 2018.

To Whom Is It Relevant? What Gaps Does This Workforce Construct Aim to Fill?

This construct is focused on the supply side of the equation and could be applied to *any* shortage specialty if and when technology becomes available to make it feasible under remote conditions. In addition to facilitating RC participation among broad populations that are underutilized in existing RC workforce constructs, the Telereserves model could be used to draw talent into several in-demand specialties identified in Chapter Three. IT and cybersecurity work, for example, may be more likely to involve tasks that can be performed remotely and independently using technology than work that is more hands-on or requires a greater degree of collaboration. Several state and local government workforce experts we interviewed cited IT jobs as more suitable to a telework model than jobs that involved regular interaction with the public. Moreover, technology companies in the private sector are well represented on lists of companies that are telework friendly.[14] Not only are IT tasks potentially more conducive to being done remotely but also, because of the stiff competition for skilled technology professionals in the public and private sectors, introducing such flexibilities as expanded remote work opportunities could be a valuable way to tip the balance for these individuals in the direction of devoting some of their valuable time to the RC.

Beyond expanding opportunities to telework—RC members essentially doing the same things they would do at their duty station but with greater flexibility to do them remotely—the Telereserves model also imagines the possibility of using advances in technology to push the boundaries of what is considered a task that requires in-person interaction and what can be done remotely. One area ripe for further exploration of remote work models is in the realm of telemedicine, which has the benefit of overlapping with the medical specialty shortage identified in Chapter Three and highlighted in a recent GAO report.[15] Telemedicine (also known as telehealth) is in use in the pri-

---

[14] Laura Shin, "Work from Home 2018: The Top 100 Companies for Remote Jobs," *Forbes*, January 17, 2018.

[15] GAO, 2018a.

vate sector, as discussed in Chapter Six, and telemedicine is reimbursable through public health care programs, including limited services through TRICARE.[16]

To be sure, the Military Health System has implemented elements of telehealth in recent years (though just 1 percent of active-duty service members utilized telehealth in FY 2016 and a small subset of military treatment facilities accounted for almost all of these encounters)[17] and sought to expand these programs and document their efforts (in line with a requirement in the FY 2017 NDAA).[18] However, most of these programs appear to be designed with the goal of expanding access for patients rather than enhancing flexibilities for providers to work remotely themselves. To cite language used to describe behavioral health telemedicine, "These encounters are typically delivered from provider hubs to various patient 'spoke' sites."[19]

Imagine, rather, if the providers themselves were at "spoke" sites, perhaps so distributed such that each spoke was the medical professional's own private practice or even their home. This could help to facilitate RC participation of those with geographic or scheduling obstacles to traditional service, such as medical professionals who live in rural or remote areas and are unable or unwilling to travel long distances for drill weekends and annual training; those who work weekends and therefore have schedules that are incompatible with the traditional RC model; those who have their own medical practices and are unable to be away for an extended period; and, potentially, those who are older than traditional RC age and/or are semiretired if the Telereserves model were to be combined with the No Passport Required model that does not require members to be deployable. A May 2018 report by the Institute for Defense Analyses found "the standard Reserve arrangement

---

[16] TRICARE, "Covered Services: Telemedicine," webpage, undated(a).

[17] GAO, "Department of Defense: Telehealth Use in Fiscal Year 2016," memorandum to congressional committees, GAO-18-108R, November 14, 2017c.

[18] DoD, *Enhancement of Use of Telehealth Services in the Military Health System*, report in response to Section 718 of the National Defense Authorization Act for Fiscal Year 2017 (Public Law 114-328), October 2017.

[19] DoD, 2017.

(one weekend a month and two weeks a year) is poorly suited for high-skill medical professionals (e.g., high opportunity cost from impact on private practice) and they are generally not efficiently utilized during these periods (e.g., they perform medical administrative work, routine medical care for Reservists, and backfill low-volume MTFs)."[20] Plugging these high-skill professionals into a Telereserves model is one possible way to change the equation—to apply their knowledge and skills to harder-to-reach cases while lessening the burden placed on them.

Among the other fields that could benefit from a Telereserves model are linguists, another shortage specialty identified in Chapter Three. Whether by expanding a traditional telework model whereby linguists perform the same tasks they would perform at their duty station but in another location or by considering how new technologies could shift the line between what is considered work that can and cannot be done remotely (e.g., doing a wider array of translation work remotely), there may be space to enhance the geographic flexibilities for these individuals. Moreover, the pool of talent with relevant language skills could overlap with the population of people that face particular location-based obstacles to RC service, for example, overseas language teachers flagged in the two-by-two rubric presented in Chapter Three.

Potential Disadvantages

A recent RAND report identified four potential costs to employers from telework programs, as generally practiced: (1) direct costs associated with fielding the infrastructure that makes telework possible; (2) potential direct costs of losing proprietary information, secrets, and data; (3) potential indirect costs associated with loss of line-of-sight supervisory control of employees; and (4) potential costs in loss of the innovation and agility from ad hoc employee interactions.[21] Pending design of a specific construct, these can all be listed as potential areas of risk.

---

[20] Whitley et al., 2018.

[21] Weinbaum, Triezenberg, et al., 2018.

From a military standpoint, collective participation in an activity achieves at least two objectives—not only does the individual participant learn or refresh whatever skills are directly involved, but also leaders in the group gain experience training and/or supervising the other participants in the process. While telework will create its own requirements and opportunities for leadership and management from a distance, it will also take away some of benefits that personal contact has on developing these leaders.

Workforce Construct 3: Reserves on Demand

What Is This Workforce Construct?

This option envisions the development of a technology platform, including a smartphone application, to match RC members with the requisite knowledge, skills, and availability with service opportunities of varied durations quickly and efficiently. The Reserves on Demand model could be a natural complement to the Telereserves model, which in tandem would facilitate making contributions when and where it fits individuals' schedules. It could also be paired with the Part-Time Plus model, allowing individuals who already serve in the RC in a more traditional capacity to supplement their service by stepping in to accomplish tasks as they arise, or with the No Passport Required model, for on-demand service opportunities within the continental United States. Like those models, this one is supply focused and is most applicable to increasing the utilization of individuals already in the Selected Reserve rather than recruiting new RC members.

If a smartphone application were used to solicit volunteers and match RC members with traditional deployments, all manner of requirements could be met with this model. However, to the extent Reserves on Demand is used to match individuals with short-term tasks or with remote work, the range of tasks that would mesh well with the model would be limited. Work that can easily transfer from person to person over time or that can be divided into discrete tasks may be better suited for an on-demand model than other tasks, and work to be done remotely would face the same obstacles discussed above related to the Telereserves model.

An existing model in use in the RC that could be considered an early-stage precursor to a Reserves on Demand model is the World-wide Individual Augmentation System (WIAS), which is designed "to manage individual augmentation requirements, sourcing, and account-ability" and provides a mechanism to match requests for additional manpower with individuals to fill these positions.[22] In some cases, Army commands, Army service component commands, and direct reporting units are tasked with identifying individuals to fill gaps, and they can do so in part by soliciting volunteers from among the RC.[23] In other cases, U.S. Army Forces Command itself, which manages the WIAS tasking process, solicits volunteers for overseas missions and matches them with opportunities based on their experience and skills.[24] While these opportunities are typically longer-term deployments, the general mechanism—a centralized system to match volunteers who have time to serve with opportunities to do so that align with their skills—would be consistent between WIAS and a Reserves on Demand model.

More technologically advanced, close-to-real-time platforms for matching individuals with time and skills to contribute with organiza-tions with work that needs to be done include the applications devel-oped and in use in the private sector, including Upwork and TaskRab-bit. As discussed in Chapter Four, a number of large corporations are using these mechanisms to offer flexible surge capacity to their work-force and to draw in skills that may not be readily available in-house. Some state and local governments have turned to these gig economy platforms as well, though several interviewees noted that there can

---

[22] National Archives and Records Administration, "Request for Records Disposition Authority," Worldwide Individual Augmentation System, Records Schedule No. DAA-AU-2016-0036, September 20, 2016.

[23] Deputy Chief of Staff of the Army, G-1, "Army Mobilization and Deployment Reference (AMDR)," last updated January 18, 2018.

[24] U.S. Army Human Resources Command, "Volunteering for Worldwide Individual Aug-mentation System (WIAS) Positions," webpage, last updated May 14, 2018.

be impediments to using this form of flexibility in the public sector, including labor regulations and budgetary requirements.

Whom Does This Workforce Construct Target?

The Reserves on Demand model would be geared toward reducing the scheduling obstacles identified in Chapter Four that impede RC service among a range of populations based on their personal characteristics or the nature of their civilian occupations. This would be accomplished by making the terms of that service more flexible, allowing individuals to volunteer to serve at times that work better for them. Individuals with small or unpredictable chunks of time to contribute, including existing RC members who could use this mechanism to perform additional service, are among the populations that could be utilized more intensively through this model.

Single parents may also benefit from this model, if it were combined with the Telereserves model, since it could allow them to serve at odd hours. An interviewee from an organization that conducts work related to military families stated that Reserves on Demand seems to make sense and mentioned, "[W]hen the kids are asleep is a great time to work as a single parent."[25]

A differentiator between the use of a platform technology to source workers in the private sector or public sector outside of the military and the proposed Reserves on Demand model is that the labor pool accessible via the platform would be restricted to individuals that are members of the RC—and the set of tasks seeking workers publicized and filled using the platform would likewise be restricted to the RC. This would bypass many of the concerns about using these platforms in other environments, such as their implications for labor standards (e.g., are these workers contractors or employees?) and challenges in verifying whether individuals have the relevant skills to meet the organization's needs.

---

[25] Telephone interview with representatives from an organization that focuses on issues related to military families, July 18, 2018.

To Whom Is It Relevant? What Gaps Does This Workforce Construct Aim to Fill?

Some roles filled through this mechanism could be more traditional in nature, involving a commitment of a substantial block of time at a fixed location. In other cases, this model could be used to allow RC members to pitch in during spare hours that fit their schedule, picking up the slack when surge capacity to accomplish discrete, time-limited tasks is needed. It would apply to a wide variety of shortage specialties.

Potential Disadvantages

Several of the obstacles present in other sectors would carry over to RC implementation of a Reserves on Demand model, including how to integrate on-demand RC members completing discrete tasks with more traditional RC and active-duty members, as well as the implications for pay and benefits when individuals serve for amounts of time that may differ substantially from the pay periods to which the RC is accustomed. In addition, there could be technological and cultural obstacles to developing a smartphone application for sourcing RC tasks, as well as challenges in integrating an on-demand, volunteer-driven model into existing mechanisms for matching individuals with needs. Nonetheless, as technology advances and with younger generations more fluent in it, not to mention more drawn to flexible work models, a Reserves on Demand model could help the RC keep up with changes in the broader economy.

Adopting this type of nonstandard work arrangement includes such risks as potentially leading to more turnover and a need to continually onboard and train workers, increasing "coordination and integration costs," and using such work arrangements may be a sign of a "lack of commitment to the workforce."[26] Specifically, performance management may be a challenge, particularly for military leaders used

---

[26] Elizabeth George and Prithviraj Chattopadhyay, "Understanding Nonstandard Work Arrangements: Using Research to Inform Practice," Society for Human Resource Management and Society for Industrial and Organizational Psychology, 2017.

to direct contact with subordinates.[27] Another possible disadvantage may emerge as active-duty members may feel left out and may want to "bid" on opportunities. In other words, the AC may resent that it is not accommodated with such flexibility if AC members witness such flexibility in the RC. If extreme flexibility is applied to members of the force because they otherwise would be unable or unwilling to participate, many RC members who can participate in the more traditional models may want the same flexibility. See the section "Potential Benefits and Drawbacks of Nonstandard Work Arrangements" in Appendix C for a more complete discussion of these risks.

Workforce Construct 4: Seasonal Worker, Seasonal Reserve

What Is This Workforce Construct?

A particularly intriguing construct combines most clearly the potential for military flexibility to both mirror and take advantage of changes in the civilian marketplace. This construct entails a system whereby reservists can commit a continuous month or more per year to military training or employment and is aimed at capturing the segment of the population with seasonal civilian jobs, while also providing a block of manpower to meet spikes in military requirements or train in specialties with high "start-up" costs.

This service option would involve a new conceptualization of the training requirement such that individuals unable to commit a regular weekend a month and two weeks a year could participate in the RC less evenly throughout the year, serving at higher levels of participation during some periods and not serving regularly in others.

This suggests developing a new form of service that (1) maintains the same total service day requirement (39) but provides more flexibility in when those training days occur (i.e., monthly participation is

---

[27] For a discussion of the private-sector context, see Jon Younger and Norm Smallwood, "Performance Management in the Gig Economy," *Harvard Business Review*, January 11, 2016.

not required as long as the total day requirement is met) or (2) expands the total number of service days but provides more flexibility in when those days occur (i.e., participants could serve more than the current minimum days but concentrated in a period that may exclude several months at a time). While its use for the latter purpose could lead to increased total manpower spending for the RC, this would only occur when there was a funding source that allowed for such extended training. Presumably, this would combine the training value with a substantive mission (e.g., the socioeconomic impact of repairing a road in a remote area or strategic benefit of supporting an overseas exercise).

Whom Does This Workforce Construct Target?

A large number of individuals in the population are potentially unable to meet the monthly RC training requirements for portions of the year due to block-month or seasonal labor force participation but may have the capacity to participate at higher than minimum requirement levels of engagement during other segments of the year. In the ACS data, 11 percent of all workers reported having predictable seasonal fluctuations in their work schedule, going to 24 percent for farm managers and 22 percent for transportation, construction, mechanics, mining, and agriculture workers. (See Table B.1 in Appendix B.) Another source, 2014 BLS data, indicates that about 5.92 million people are employed in highly seasonal industries—about 4 percent of the total employed population in 2014.[28] The requirement to participate in training exercises one weekend a month, every month, for example, may present either an insurmountable obstacle for some individuals or provide strong disincentives to enlist.

Thus, one approach to making service in the RC more feasible for this population is to develop service schedules that take blocks of unavailable time into account, while also potentially capitalizing on additional service amount opportunities in the available blocks—Seasonal Worker, Seasonal Reserve.

---

[28] Breakdowns of specific occupations with winter and summer peaks can be found in BLS, "Industries with Winter Employment Peaks," *Economics Daily*, December 3, 2014b; and BLS, 2014a.

Among the civilian occupations that seem particularly promising under this model are agricultural workers, tax preparation workers, those in industries with summer employment peaks, including hotels and motels, fitness and recreational sports centers, golf courses and country clubs, amusement parks and arcades, museums, historical sites and zoos, marinas, and recreational and vacation camps. Winter employment peak industries include temporary help services, many categories of retail stores, electronic shopping, and skiing facilities. Other occupations, such as commercial fishermen or overseas language teachers or postsecondary students, may require uninterrupted months of work at various periods throughout the year, possibly at different locations, or locations that are not near training bases.

The population to which this applies does not have a specific or narrow set of skills, as seasonal work spans a wide range of occupations. However, the broad nature of this underutilized population likely contains an array of specifically sought-after skills, such as training in science, technology, engineering, and mathematics (STEM) fields or linguistics, in addition to many others.

To Whom Is It Relevant? What Gaps Does This Workforce Construct Aim to Fill?

Construction and transportation are two examples of shortage specialties that are hard to meaningfully train in a 36-hour battle assembly but—if grouped into a month of sustained collective training and activity—could not only build individual and collective skills but also produce outputs of value to military or civilian communities.

Potential Disadvantages

Several challenges to the development of this type of service include the need to coordinate the participants to ensure enough people are present for drills at the same time. This could involve temporarily relocating participants to bases farther than their usual/current training location for the duration of their training period. It also would require changes to payroll programs and other automated processes to account for the irregular training periods. This also may involve the reconceptualization of the role of unit cohesion for RC participants (e.g., if they do not all regularly train at the same time in the same location, they may not form the same bonds as current RC participants). Some occupations may not be appropriate for a form of service that potentially involves several months devoid of training; cyber skills, for example, must be kept current.

The most commonly noted risk associated with this construct is the potential for individual and collective skills to atrophy over months without training. While a construction team might be battle ready after a month of working together, six months later individuals may have moved to new units and the rest lost that acquired teamwork. We consider the pace of this "team decay" to be an empirical question worth assessment in a workforce construct, not a reason to avoid experimentation.

Workforce Construct 5: Job Sharing

What Is This Workforce Construct?

One approach to increasing RC capacity, focused on increasing the amount of time served rather than increasing the number of participants, could be developed based on the concept of job sharing that was discussed in Chapter Four. This option could include job sharing between two (or possibly more) people, as well as shift trading options, that would in many ways effectively be similar to part-time working arrangements that would combine across individuals to replace a single full-time position. Individuals could thus work across the part-time spectrum from low commitment to nearly full time, depending on the time split.

A range of businesses and government offices offer job sharing as a flexible work option. The Church of England has a job-share program for vicars, and there are job shares in the Supreme Court of the United Kingdom.[29] One interviewee with extensive HR experience in both the private and public sectors noted that job sharing was popular,

---

[29] Gaby Hinsliff, "Judges, Soldiers, MPs, Vicars—Can Job-Sharing Work in Any Field?" *The Guardian,* December 2, 2016.

more so in the private sector, particularly for people who were retiring younger and healthier and who wanted to remain active but not with a full-time position. This may analogously appeal to the Individual Ready Reserve, as noted.

As noted previously, the Australian military offers job sharing with another employee as one of its work-life balance options. In the Australian example, "the two employees are employed on a part-time basis and combine to perform a job normally filled by one person working full-time."[30] Australia's SERCAT 6 classification effectively incorporates job share for permanent members who wish to render a pattern of service other than full time. The RC of the Australian military does not qualify for SERCAT 6, but extending a similar model to the U.S. RC to facilitate greater participation by members may be appealing and effective in staffing what are effectively full-time positions across individuals.

Whom Does This Workforce Construct Target?

Individuals may be interested in taking advantage of job-sharing arrangements to participate in the RC at greater than minimum levels, despite being unable or unwilling to manage the demands of a full-time post. Former active members who wish to continue to serve but would prefer a more part-time schedule would also be a potential target for this option. This option could also appeal to current RC members who want to keep their specific role but want more flexibility. Thus, the program could capitalize on the specific skills of current and former RC and AC participants through increased time spent in RC activities. As noted by an interviewee from an organization that focuses on issues related to women in the military, there may be an untapped population who wish to transfer from the Individual Ready Reserve to positions providing more commitment. They would usually still have clearance,

---

[30] Australian Government Department of Defence, "Work Life Balance," webpage, undated(c).

may have more school, already meet height and weight requirements, and maintain online courses to obtain their points for the year.[31]

To Whom Is It Relevant? What Gaps Does This Workforce Construct Aim to Fill?

This is particularly demand focused, as it would apply most commonly to missions conducted continuously using a specific platform, whether an airframe or a desk in a 24/7 intelligence watch center. A number of shortage specialties fall under this kind of employment model and therefore are candidates for this construct. Further, job sharing may be one method of reducing burnout and attrition among high-stress positions. In the RC context, job-sharing arrangements where two (or possibly more) people share the same workload and divide up the days (typically referred to as the "twins model") would most likely be more appropriate than other types of job share in which people split one job into discrete tasks for which they hold responsibility (the "islands model" or "job split"), although both are found in the civilian workplace.

Potential Disadvantages

RAND identified both financial and administrative risks associated with this construct. Not only would DoD need to develop IT and HR systems to track, manage, and compensate personnel performing these duties, but eventually it also would raise questions about long-term compensation. By falling somewhere between full- and traditional part-time duty, this construct could create a class of individuals with either unusually high numbers of duty days credited toward their reserve retirements or enough days of active duty to argue for a regular retirement. Given that these individuals would largely have been filling critical, long-standing needs that could not be met through any other acceptable means, RAND expects that DoD policymakers can anticipate this challenge and plan appropriate responses.

---

[31] Telephone interview with representatives from an organization that focuses on issues related to women in the military, July 17, 2018.

Workforce Construct 6: Part-Time Plus

What Is This Workforce Construct?

The Part-Time Plus construct would focus on making it easier for reservists to serve more than their statutory minimum training days when their service has a need for additional manpower. This would provide more flexibility for those who want the benefits and stability of RC service. It would target missions and requirements that demand more than the traditional reserve duty expectations and the service members who may be interested in serving more frequently but less than full time. As such, it is a supply-side model and could be used to meet requirements across a large number of shortage specialties. It is particularly suited to skills that require higher usage to maintain credentialing.

This model would require the services to create a competitive benefits package for service members to select from. For example, one option could be that service members who are in service a minimum of 120 days per year would be eligible for a lower premium health plan or increased contribution to the Thrift Savings Plan. A key enabler for this model would be dedicated and programmed funds to support increased RC utilization. Finding funds in years of execution would limit the potential success.

Whom Does This Workforce Construct Target?

This model seeks to primarily engage with high-demand, highly skilled service members, particularly in specialized, niche fields, although broader applications may apply.

It would provide a more flexible option for those who want the stability of a job that provides steady income, the option for affordable insurance, and meaningful work while simultaneously providing flexibility to be able to start a business, take care of children or aging parents, pursue additional education, or meet some other need in their life. Many serving in this model may choose to not pursue a civilian career. A service member would be able to say that they will commit to working three days a week but need to spend the other two days in class, with family members, developing their business, or pursuing some other goal. There is evidence in the literature that highly talented and highly skilled employees are pursuing self-employment to provide this flexibility for themselves.

To Whom Is It Relevant? What Gaps Does This Workforce Construct Aim to Fill?

This model would potentially appeal to a wide range of people but will focus on those with high-demand, high-value skills, such as aviation workers, linguists, medical specialists, and cyber warriors. Indeed, with the high demand for aviation talent and reported shortages across the services, this option may help to retain pilots and crew members who want to continue to serve but may prefer a slightly slower pace or the opportunity to homestead in one place. Cyber warriors, especially those with elite skills, need to meet high currency standards and have ample time learning the systems they are operating. In a more traditional field, a military medical facility may not have a consistent need for a particular medical specialty, and the risk of not having someone immediately available is relatively small. Regular duty days could be arranged for the niche specialist, within existing scheduling processes, and the RC physician to meet this level of demand. The Part-Time Plus model would be highly suitable to skills that require higher usage to maintain credentialing. Any military requirement with a high demand to meet steady-state requirements may consider employing this model.

Potential Disadvantages

As with the job-sharing construct, the major risk associated with this model is in its development of unrealistic expectations: if this program offers a range of desirable benefits, many RC members may apply, and at some point supply may exceed demand. This risk may be mitigated through planning and clear communication.

Workforce Construct 7: Sponsored Reserve

What Is This Workforce Construct?

This option addresses the issues surrounding the deployment of service contractors that have emerged over the past 17 years of conflict and that may be even more acute in future conflicts with near-peer adversaries. These issues include the assurance of adequate training in self-defense, survivability, integration of service contractor elements into military forces, compliance with military orders and law, and provision of long-term health and benefits support. In this model, DoD would adopt the British model of sponsored reserves (see Chapter Five) for service contractors whom it plans to deploy alongside military forces in active areas of armed conflict, where there is a reasonably high likelihood of exposure to combat.

Under this model, these service contractor employees would simultaneously work for their civilian employers and serve as reservists. In practice, a task order to their service contractor employer would also function as a mobilization order for them, bringing them into an active status for purposes of work and deployment. The precise parameters of work and service would need to be defined, however, including pay, benefits, insurance coverage, legal liability, and other issues.

This program will help DoD manage some of the legal, operational, and long-term care challenges that have arisen during 17 years of war as it has made extensive use of service contractors on the battlefield. Currently, DoD contracts with firms that deploy their employees to active areas of conflict. The terms of their service are defined by private employment agreements as well as by the government's contracts with their firms, including a number of regulations specifically applicable to contractors' service overseas. Historically, contracting firms have been called upon to deploy tens of thousands of their employees to support the conflicts in Iraq and Afghanistan, as well as many thousands more host-nation and third-country nationals. However, a number of issues have arisen during these deployments that may be addressed through a sponsored reserve model, including

- *Legal status.* The legal status of civilian contractors accompanying the force (as combatants or noncombatants) is unclear; so, too, is the extent of their combatant privilege for the use of force. This may affect the extent to which they may be lawfully targeted or their conditions of detention in future conflicts. Under a sponsored reserve program, these personnel would be combatants under the Third Geneva Convention, with no ambiguity regarding their legal status.
- *Operational control.* Under the terms of their government contracts and employee agreements, contractor personnel work for their firms, which, in turn, work for the government—specifically, for the contracting officer empowered by law to direct performance and allocate funding. In practice, contractor personnel often work more closely with military forces, but the formalities of government contracts often impede this ability. They also do not have the same legal obligations with respect to obedience of lawful orders nor the same accountability for misconduct under the Uniform Code of Military Justice. Sponsored reserves would be legally accountable to their military chains of command under the code, which may ease some issues related to operational control and discipline.
- *Long-term care and support.* Contractor personnel who sustain injuries or illnesses while performing their duties may obtain long-term support under the programs established by the Defense Base Act, which is effectively a government-subsidized workers' compensation scheme for war-zone contractors. However, this system does not provide comprehensive support comparable to what VA provides to service members. As sponsored reserves, these personnel would earn entitlement to VA benefits as part of their service, including VA disability compensation and health care.

To some extent, the MilTech program is a variant of this concept, wherein MilTechs work in a civilian role until they are mobilized and deployed with their reserve units in uniform. Similarly, there are government civilians who deploy with their organizations, but they do not convert into military personnel when they do so. A third existing form of this relationship is the U.S. Navy's Strategic Sealift Midshipman Program, which offers Navy Reserve commissions to students at the seven merchant marine academies who also complete Navy ROTC courses and meet other requirements. These individuals are then commissioned USNR Strategic Sealift Officers and serve in the IRR for a minimum of eight years.[32]

A sponsored reserve program would likely require changes to several sections of 10 U.S.C. to create specific authorities allowing for their organization, compensation, status, and mobilization. These authorities would then need to be operationalized through modifications to existing defense contracts or terms in new contract solicitations. Government contractor employees would need to be incentivized to join the sponsored reserve program, or contractors would need to be incentivized to recruit current reservists as employees, or some combination of the two.

Whom Does This Workforce Construct Target?

In the U.S. Total Force, sponsored reserves would target the population of service contractor employees who currently support DoD in areas of active conflict. These personnel have already displayed some degree of propensity for service by volunteering to work for contractors in work zones; some may also be veterans.

To Whom Is It Relevant? What Gaps Does This Workforce Construct Aim to Fill?

Such service contractors might provide logistics support, intelligence and linguistic support, construction services, or security. This model would enable DoD to use these service contractors for these functions while deploying them in a uniformed capacity, addressing the issues that have arisen with respect to contractors on the battlefield.

A model could be created within the functional area of a given U.S. defense contractor that is expected to accompany U.S. forces during a future contingency operation, such as one of the large construction

---

[32] See Naval Reserve Officers Training Corps, "Strategic Sealift Midshipman Program (SSMP)," webpage, undated.

firms that may be called upon to build runways or port facilities or a large logistics firm. The employees of this firm would work on DoD contracts in nonconflict areas, just like any other employees. After the contractor mobilizes for an overseas task in an area of active conflict, the company's sponsored reserves would also mobilize to serve in a uniformed capacity, equipped with military equipment (which would have the status of government-furnished equipment under the contract), and task organized as part of the military command structure. Under the British model, defense firms continue to pay their sponsored reserves during their deployments, billing the government for these amounts. In a U.S. model, it may be possible to mimic this practice or have DoD assume responsibility for compensation and benefits associated with active service by sponsored reserves.

Potential Disadvantages

Depending on the specifics of these legislative changes, there may be risks to existing patterns of recruitment and HR management. To the degree that existing sources of manpower opt for Sponsored Reserve positions and not ones in the broader RC, where they can learn a wider range of skills and serve in more varied positions, this might reduce the pool of individuals qualified for more senior leadership positions. Identifying this risk is the first step toward assessing its likelihood and designing policies targeted at mitigation.

Workforce Construct 8: Wounded Warriors

What Is This Workforce Construct?

This option focuses on a defined population more than a shortage requirement or specific programmatic change. The premise is that many individuals who suffered career-ending injuries or illnesses as a result of military service presumably combine valuable experience from their time in uniform and challenges in transferring them to civilian employment. To the degree their infirmities limit their full-time employment options outside the military too, part-time and flexible employment in a military-support capacity might be preferable to them as well. Unlike the other models, this does not assume that direct salaried compensation is necessarily involved. Building on the examples we will review, it is possible that veterans may volunteer for such a workforce construct in return for the intrinsic and community value of the work or possible indirect benefits, such as access to training, civilian education, or other benefits. Because of their expected constraints on appropriate duties, this construct was initially conceived as a way to provide support to units at home stations, like the current MilTech program or the No Passport Required construct described earlier.

Several models currently in existence contain similarities to this concept. First, the military already has programs for some disabled veterans to remain on duty, either with the regular or reserve forces. (The Army's is divided into Continuation on Active Duty and Continuation on Active Reserve programs.)

Second, there are a number of organizations through which individuals provide support to the armed forces while not actually serving as a full member of the federal military. Examples include state defense forces, U.S. Coast Guard auxiliary, and Civil Air Patrol.

Internationally, this construct has similarities to the Estonian cyber force described in Chapter Five, although here the missions are defined more broadly, and the focus would initially be on a smaller population of former service members.

State defense forces (also known as state military, state guards, or state military reserves) are the closest parallel to the U.S. Coast Guard Auxiliary and Civil Air Patrol for land forces. These volunteer organizations are almost entirely state assets and have a particularly strong relationship with their respective National Guard membership and leadership. Their roles vary by state, with some very active and operational and others functioning more in supporting roles.[33] For example, the Maryland Defense Force lists a medical regiment, engineer regiment, band, chaplain corps, financial corps, IT and communications

---

[33] Eric Kelderman, "State Defense Forces Grow, Project New Image," *Stateline*, December 31, 2003.

directorate, cybersecurity unit, judge advocate general corps, and troop command among its formal units.[34]

The U.S. Coast Guard Auxiliary has three missions: (1) to promote and improve recreational boating safety; (2) to provide trained crews and facilities to augment the Coast Guard and enhance safety and security of our ports, waterways, and coastal regions; and (3) to support Coast Guard operational, administrative, and logistical requirements. The last of these is the most relevant to this model. According to its official website, the specific capabilities of the U.S. Coast Guard Auxiliary include the following:

- safety and security patrols
- search and rescue
- mass casualties or disasters
- pollution response and patrols
- homeland security
- recreational boating safety
- commercial fishing and vessel exams
- platforms for boarding parties
- recruiting for all Coast Guard services.

In addition to the above, the U.S. Coast Guard Auxiliary operates in any mission as directed by the commandant of the U.S. Coast Guard or Secretary of Homeland Security.[35]

The Civil Air Patrol's relevance to this study is highlighted in the following sentence from its website: "When Civil Air Patrol formed in the early days of World War II, many of our first volunteers were patriotic citizens unavailable for military service who nevertheless were determined to serve the nation in a time of need."

The Civil Air Patrol describes itself as "a private, non-profit humanitarian organization, which is the official auxiliary of the United States Air Force." Its volunteers support emergency services and operations (serving as aircrew members, communications personnel, admin-

---

[34] Maryland Defense Force, "MDDF Units," webpage, undated(b).

[35] U.S. Coast Guard Auxiliary, "About the Auxiliary," webpage, undated.

istrative staff, financial managers, logistics and supply personnel), its cadet program (mentoring more than 23,000 cadets in leadership, aerospace education, physical fitness, and moral and ethical decision-making), and its aerospace education program (providing training and resources to teachers). Volunteers are unpaid and pay annual dues and uniform costs to participate. The Civil Air Patrol explicitly encourages current and former armed forces officers and NCOs to join.[36]

A final relevant model is the Estonian cyber corps discussed in Chapter Five. As noted there, this is an option that particularly caters to those seeking a "less military" form of national service. As it develops, it may provide insights into the advantages and disadvantages of using volunteers to perform sensitive tasks, such as cyber defense; if successful, it might be one way to creatively meet this growing DoD requirement.

Given these precedents, the challenge is to create a program that would leverage the experience of wounded warriors without competing unnecessarily with the existing programs. A program that would support the federal military through its federal RCs and fall outside of the air and sea transportation and emergency response functions already covered by the Civil Air Patrol and U.S. Coast Guard Auxiliary would seem to fill the gap in this regard—in other words, a Federal Military Support Corps, whose members would wear a modified military uniform and provide voluntary support to Army, Air Force, Navy, and Marine Reserve units at or near their homes. They could augment the unit's FTS staff and chain of command by helping plan and execute training (e.g., helping write training plans and serving as mentors or role players) and assisting in routine personnel and maintenance activities. DoD would have to decide how formally to manage these tasks. The longer the list, the more questions will arise, such as

---

[36] Active, retired, or honorably discharged personnel may be advanced to a Civil Air Patrol grade equivalent to their grade in the armed forces (but not to exceed lieutenant colonel) in recognition of their military knowledge and experience. Regular, reserve, and National Guard senior NCOs, active or retired, in the grade of E-7 through E-9 (including those in the Coast Guard), may receive an advanced grade in recognition of their military knowledge and experience. U.S. Air Force Auxiliary, Civil Air Patrol, "Who We Are," webpage, undated.

whether Federal Military Support Corps members would be issued Common Access Cards (required to support many HR functions) and how they would be insured (necessary if they will use equipment in a motor pool or in the field). While this concept could be expanded to include all types of willing individuals, one could argue that it should begin with and focus on seriously wounded warriors, as their military experience and presumed inability to serve in the RCs make them an ideal fit. Assuming the Federal Military Support Corps remains an unpaid body, the government or others could provide indirect benefits, including preferential opportunities for education, training, or paid positions.

Whom Does This Workforce Construct Target?

While estimates vary, the number of individuals who are (a) veterans, (b) able to work at least part time but (c) disabled to such a degree that they are not ideal targets for RC recruiting probably reach into the hundreds of thousands. Among the things known about this group are the following:

1. This is a large segment of U.S. society. More than 3 million service members have deployed since 9/11.[37] In its most recent annual report on disability compensation, as depicted in Table 7.2, VA reported 1,060,408 post-9/11 Global War on Terror veterans receiving disability compensation.[38] The median number of service-connected disabilities for post-9/11 veterans is 7.44 (as compared with 2.41 for World War II veterans, 3.78 for Vietnam-era veterans, and 3.30 for peacetime veterans). Of the 1,060,408 post-9/11 veterans drawing VA disability compensation, 682,291 do so with combined ratings of 50 percent or higher; 111,703

---

[37] According to data from the Defense Manpower Data Center's Contingency Tracking System; see also Jennie W. Wenger, Caolionn O'Connell, and Linda Cottrell, *Examination of Recent Deployment Experience Across the Services and Components*, Santa Monica, Calif.: RAND Corporation, RR-1928-A, 2018.

[38] VA, *Veterans Benefits Administration Annual Benefits Report, Fiscal Year 2016*, Washington, D.C., 2017.

Table 7.2

Veterans Health Administration-Reported Data on Global War on Terror Compensation Recipients

Combined Degree (%)	Male		Female		Total	
	Number	%	Number	%	Number	%
0	527	0.06	85	0.06	619	0.06
10	94,579	10.37	13,152	9.17	108,672	10.25
20	66,162	7.26	9,637	6.72	76,497	7.21
30	78,787	8.64	12,836	8.95	92,092	8.68
40	85,731	9.40	14,088	9.82	100,237	9.45
50	72,209	7.92	12,182	8.49	84,770	7.99
60	111,202	12.20	16,049	11.19	127,777	12.05
70	105,372	11.56	16,187	11.29	122,010	11.51
80	113,870	12.49	17,858	12.45	132,260	12.47
90	88,132	9.67	15,282	10.66	103,771	9.79
100	95,232	10.44	16,063	11.20	111,703	10.53
Total	911,803	100	143,419	100	1,060,408	100

SOURCE: VA, 2017, p. 22.

have combined ratings of 100 percent from VA. Alongside these veterans drawing VA disability compensation, there is a parallel population of DoD disability retirees drawing a medical retirement from DoD. As of September 30, 2017, there were 118,029 DoD disability retirees, who had an average age at the time of 58.7 for officers and 50.7 for enlisted personnel.[39]

2. Disabled veterans desire flexibility. In a literature review conducted by the National Organization on Disability, "[c]apacity for flexibility, such as working hours and working environment"

---

[39] DoD, Office of the Actuary, *Statistical Report on the Military Retirement System, Fiscal Year 2016*, July 2017.

was the top trend in veteran preferences and support need for employment.[40]

3. Seriously wounded veterans are often still supporters of the military and other service members. As shown in Figure 7.1, Pew research found that almost all seriously injured veterans felt proud of their service, and more than two-thirds would encourage others to enlist.

4. Many existing options try to capture the potential of veterans by prioritizing their application for government employment[41] or giving a preference to firms owned by disabled veterans in government contracting selections.[42] However, these may not be the best options in all cases, particularly for seriously injured or ill veterans who require more flexibility to accommodate their disabilities or want to remain more directly connected with national security.

5. As with some of the other models, such as No Passport Required, Wounded Warriors could provide valuable service by augmenting the FTS personnel at most units. While many FTS duties require specific knowledge or position authorities (such as personnel actions requiring approval from the chain of command), many are administrative and logistical. For example, there's a fixed cost from simply having a reserve center open for business—someone has to unlock the building, answer the phone, take information from a potential walk-in recruit, or sign for deliveries. All these could be done by a volunteer or part-time employee who "knows the system." If "hired" early after their separation, any active security clearances or certifications for

---

[40] National Organization on Disability, *Return to Careers*, New York, November 2011, pp. 73–74.

[41] Feds Hire Vets, "Special Authorities for Hiring Veterans," webpage, OPM, undated.

[42] See 38 U.S.C. 8127, Small Business Concerns Owned and Controlled by Veterans: Contracting Goals and Preferences, January 3, 2012; see also U.S. General Services Administration, Federal Acquisition Regulation Subpart 19.14, Service-Disabled Veteran-Owned Small Business Procurement Program.

Figure 7.1
Injured Veterans' Attitudes

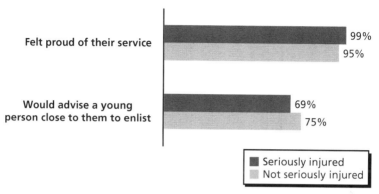

SOURCE: Rich Morin, "For Many Injured Veterans, a Lifetime of Consequences," Pew Research Center, November 8, 2011. Used with permission.

these veterans could also be preserved and retained for the benefit of the RC.

To Whom Is It Relevant? What Gaps Does This Workforce Construct Aim to Fill?

It is critical to note that this program will not directly target the shortage specialties previously identified; however, by adding to the total manpower pool, this construct may allow the military to better focus paid manpower on shortage areas.

Potential Disadvantages

Such a program may carry security risks in terms of potential abuse by individuals who might use its terms to come into closer contact with the military for some ulterior motive. Application and periodic review procedures would have to be established to screen out everyone from terrorists seeking a soft target to those who would use Federal Military Support Corps participation as a cover or tool for criminal activity against civilians.

Workforce Construct 9: Warrant Officer–Deacons

What Is This Workforce Construct?

Another way to address shortages is to redefine the requirement and therefore expand the pool of people able to meet part of the mission. We offer an example using the chaplain career field, as this is often cited as experiencing shortages. While this workforce construct is elaborated as being relevant to only one specialty and therefore is somewhat distinct from the other constructs detailed above, this approach could be applied to any specialty where the existing definition and force structure choices may be precluding options. For example, the Army broadened its pool of potential pilots decades ago when it created warrant-officer pilot MOSs, a path not taken by the other services. Because any effort to make such changes will require in-depth analysis of many facets of the specialty, we will go into some detail in describing our illustrative concept of creating a new warrant-officer career field of deacons, who would reduce the number of commissioned-officer chaplains and expand recruitment to a population more likely to be able to train on weekends.

Many policy, regulatory, and administrative changes might be tested to see their impact on the shortage of chaplains. Some of these are service specific, like how recruiting for chaplains is organized or the balance between chaplains assigned to National Guard and reserve units and individual mobilization augmentees supporting active-duty installations. Some of the other workforce constructs in this chapter could also be leveraged to cover shortages of chaplains. Therefore, this short section will focus on one chaplain-specific strategic concept, which admittedly could only be implemented after a much more robust period of discussion, development, and decision. However, we feel this concept hits on so many of the salient issues that such discussion and debate would also help stakeholders come to terms with the need for the strategic evaluation of the chaplaincy.

The essence of the potential workforce construct would be to create a warrant officer program for deacons (abbreviated here as WO-D). This combination of an established clerical role with an existing military personnel category could enable the military to not only recruit a new demographic into military service but also address demographics and functional issues in its current force.

First, what is a deacon? The Roman Catholic Church and some Protestant denominations use the title to designate individuals authorized to perform some but not all sacraments in place of an ordained priest or minister and/or support the management of church ministries. While deacons may have attended seminary, it is not required, and only 10 percent of the Catholic Church's permanent deacons have a graduate degree in a religious field.[43] While the total number of Catholic priests in the United States dropped 37 percent between 1975 and 2017, the number of deacons grew nearly 2,000 percent, so there is now almost one deacon for every 1.4 diocesan priests (i.e., excluding those who teach at universities or perform other ministries).[44] More than ten years ago, the Canadian Army and Catholic Church created a military deacon program.[45] As of 2014, one-eighth of ordained Catholic chaplains were deacons.[46] It is our understanding that the Catholic bishops of the United States have historically opposed such an option.

Coincidentally, Pope Francis "has appointed a commission of experts to study the role of women deacons in the early church, presumably with an eye to seeing if an ordained female diaconate could be permissible today."[47] If the Catholic Church were to expand the diaconate to female and/or married persons, this would complement a program of military deacons.

[43] Mary L. Gautier and Thomas P. Gaunt, *A Portrait of the Permanent Diaconate: A Study for the U.S. Conference of Catholic Bishops, 2014–2015*, Washington, D.C.: Center for Research in the Apostolate, Georgetown University, May 2015, p. 13.

[44] Center for Applied Research in the Apostolate, "Frequently Requested Church Statistics," webpage, undated.

[45] Padre Gauthier, "Diaconal Council," trans. Padre Delisle, Ottawa: Roman Catholic Military Ordinariate of Canada, undated.

[46] The full breakdown was 35 priests (26 regular, 9 reserve), 5 deacons, and 59 pastoral associates. Michael T. Peterson, *The Reinvention of the Canadian Armed Forces Chaplaincy and the Limits of Religious Pluralism*, thesis, Wilfrid Laurier University, 2015, p. 146.

[47] Nicole Winfield, "Vatican Seeks 'Courageous' Ideas to Combat Priest Shortage," Associated Press, June 8, 2018.

On the military side, other career fields have for decades used warrant officers—highly skilled specialists who do not have the same kind of command responsibilities as many commissioned officers of equivalent seniority. A WO-D program would have the following advantages:

- At the simplest level of providing additional manpower to meet a defined set of requirements, it would potentially expand the pool of individuals substantially. Where Catholic chaplains are in short supply, for example, Catholic WO-Ds could meet some liturgical and pastoral needs and allow the commissioned chaplains to cover a wider population. Coincidentally, some of the denominations suffering most from a chaplain shortage are those with a tradition of permanent deacons, such as the Catholic Church and liturgical Protestants (the Episcopal, Lutheran, Presbyterian, and Methodist communities).
- WO-Ds could shift the work burden at the higher levels as well. At a large installation, there may be a number of chaplains, largely performing staff and management functions rather than having direct interaction with their flock. A corps of WO-Ds with experience running religious support programs could take up some of that effort, which does not require the theological training and pastoral experience of an ordained clergyperson.
- Current warrant officers are often prior-service enlisted personnel, so this would be similar to other warrant-officer tracks. Current or former military personnel who feel called to serve could go into warrant-officer status without losing their retirement benefits, and more senior individuals might find a warrant officer's status within the rank structure a comfortable one.
- Chaplains, like most other officers, currently serve most of their careers in an "up or out" officer management track. As a WO-D, they could serve as long as they wanted and focus on providing religious support at the unit or installation level without a need for senior staff experience.

Unlike many of the workforce constructs, a WO-D construct would not require legislation. Because the services already have warrant officers, it would just require one of them to work through the personal and force-structure changes to create a new specialty within that path. The greater challenge will be working with the stakeholders outside the government to build a pipeline and process for accessing and managing the individuals within their faith communities.

Whom Does This Workforce Construct Target?

The target population, individuals who have an interest in religious ministry within the military but for some reason are not willing or able to seek ordination and commissioning as a military chaplain, was not identified as a particular underserving demographic group. This construct focuses on an identified gap in manning, personnel for chaplain positions, and tries to identify a force-structure option that would shift part of the chaplains' mission burden to a new MOS with a different recruiting population.

To Whom Is It Relevant? What Gaps Does This Workforce Construct Aim to Fill?

The problem of a shortage of clergy is not unique to the military, nor is it new.[48] However, where the shortages in the past were often focused on a particular service or denomination/religion, they are now much broader. To some degree, this reflects trends in U.S. culture, but that does not address the problem. Short of a radical change in how the military provides religious support to service members and their families, the demand will remain fairly constant, and the services must decide how to meet it.

The pool of potential military chaplains shares many features with other hard-to-access populations:

1. They often work on weekends.
2. The qualifications for the occupation vary greatly in the civilian world. For clergy, some denominations have very hierarchi-

---

[48] Nationally, the number of clergy is projected to grow by 19,900 between 2016 and 2026, and 29,200 positions will be created or become vacant during that period. BLS, "Employment Projections," undated(b).

cal structures and a professional clergy, which make it easier for them to recruit or even direct clergy to serve as chaplains; others are decentralized and/or use lay leaders to provide most services, often making it harder for the military to identify and validate the qualifications of would-be chaplains. This also means differences in specific skills: some chaplain candidates have done internships providing pastoral care at a church or other location or completed clinical pastoral education (e.g., working in hospitals) prior to becoming chaplains, while others complete a graduate degree largely or entirely online.

3. They have unconventional career paths. In this case, many do not complete the civilian training to become a civilian clergyperson until age 30–49, making it less likely they will (a) choose to join the military and (b) complete the necessary 20–30 years for an ideal career (ideal both for the individual's and the institution's needs).

4. As the military becomes more diverse, those who provide support to them also need to diversify. DoD currently recognizes 221 religions;[49] the Army, for example, has chaplains from 48 identified religions or denominations.

Adding complexity to the problem, chaplains (like linguists or attorneys, for example) include many nonsubstitutable subdivisions within a single specialty. For example, a generic "Protestant" chaplain may be fine for an installation's weekly Sunday service, but specific denominations may need one of their own priests to do a baptism or confirmation rite. Some denominations may not authorize their clergy to perform same-sex weddings or marriage counseling, but a chaplain's assigned population may include service members who need such pastoral care or ministry. And, obviously, not every base will have a chaplain rabbi or imam to conduct the range of distinct services for those personnel.

---

[49] Kimberly Winston, "Defense Department Expands Its List of Recognized Religions," Religion News Service, April 21, 2017.

Another challenge facing the chaplain community is the tension between staff and religious duties. One interviewee noted that chaplains share a dilemma with other professions within the officer corps: they are recruited to provide a distinct service but as they advance are expected to be more the staff officer and less the "service provider." Senior chaplains, he noted, are promoted based on how they handle budgets, staffs, and major projects, not on their pastoral care or effective preaching.

Finally, there is an obvious demographic challenge—because only single men can be ordained as Catholic priests, only single men can be Catholic chaplains.

Potential Disadvantages

There are several potential risks to this construct: (a) opposition from different faith groups might prevent its implementation, (b) the desired populations of potential deacons might not be receptive to this new form of military and religious service, and (c) the construct might initially cause confusion as the services define the duties of the WO-Ds and service members learn to differentiate between the military and ecclesiastical duties of chaplains and deacons.

How the Constructs Could Be Used

Given these nine alternative workforce constructs, the next step would be to reduce them to a smaller set of leading ideas and then conduct additional research on them. We have provisionally assigned the constructs to three categories, in order of increasing barriers to further research and experimentation in the near term:

- Several constructs require the military to *increase comfort with technology* and would probably grow incrementally as new tools from the civilian sector become available and as military managers see advantages to adopting them. In particular, the Telereserves and Reserves on Demand constructs fall within this category.

- Several constructs are *waiting for an advocate* in that they are primarily concepts that seem to have few institutional barriers and could be implemented by a service with leadership support. In particular, the Warrant Officer–Deacon, Wounded Warrior, and Seasonal Worker, Seasonal Reservist constructs fall into this category.
- Finally, several constructs require a greater *change to culture, policy, and systems*, and creating even a pilot program for one of these constructs would require a coalition of advocates to develop the personnel systems to manage a new category of reservists, change policies and statutes, and gain buy-in from both the AC and RCs. The pilot programs' feasibility would depend on whether such changes added capacity to the Total Force and were viewed as worth the costs and effort. The No Passport Required, Job Sharing, Part-Time Plus, and Sponsored Reserves share the feature of changing more fundamental terms of the RC experience.

Regardless of the category into which a construct falls, additional research would be useful and necessary in order to define the expectations of any pilot program. Such research could include estimating the effects on recruiting, retention, and readiness. Predicted changes in these outputs would interact with cost analysis of the required change—some constructs imply an investment in technology, others a trade-off of personnel costs for AC, RC, civilian, and contract manpower.[50] Training pipelines may be changed or the demand for facilities shifted.

---

[50] While these analyses are amenable to quantitative analysis, it is not necessarily clearcut, and some of the necessary data may not be immediately available. For one discussion of the challenges in measuring these trade-offs, see Reserve Forces Policy Board, *Eliminating Major Gaps in DoD Data on the Fully-Burdened and Life-Cycle Cost of Military Personnel: Cost Elements Should be Mandated by Policy*, Final Report to the Secretary of Defense (RFPB Report FY13-02), Washington, D.C,: January 7, 2013.

Designing Pilot Programs for the Workforce Constructs

Once additional research has validated a given construct's potential for improving recruiting, retention, and/or readiness and estimated the likely costs, a service could be asked (or directed) to develop a pilot program applying the construct. While a pilot for any of these nine constructs would be designed for its specific characteristics and goals, there are some fundamental tasks that would be required for all pilot programs.

Identifying Maneuver Room and Acceptable Risk to Readiness

One of the most important currencies in military force development is authorizations, the units of force structure that count how many of a service's or component's authorized end strength will be assigned to a given MOS, pay grade, or unit. One of the first questions in designing a pilot program will be how big it can be—how many authorizations the service will commit to manning by a certain process or training in a certain way. A larger investment of authorizations would increase the potential gain from the program and may give more validity to conclusions based on the pilot; a smaller investment would limit the risks that the target population fails to respond and the authorizations are "wasted" or that the alternative is so attractive it pulls individuals from other important service options. Analysis can help a service estimate where that tipping point falls, but the decision also includes value judgments about service priorities.

Additionally, the pilot program needs to describe how the prospect of improved recruiting and retention in the long term will be balanced against readiness risk in the short term. Each service and component implementing a pilot program will have to determine whether the affected force structure will remain part of the pool available for mobilization (if in an RC) and deployable for contingencies and, if not, at what point they will be inserted back into pool or rotation schedule. The services will probably want to minimize this unavailable time, not only to maximize their capacity for contingency response but also to include in their evaluation the target population's ability to respond to these real-world demands.

Initial Assessments

In most cases, the pilot programs will not only focus on a type of shortage specialty and general target population; they will be limited to a specific specialty, geographic area, type of unit, or mode of utilization (e.g., a particular tool for telework, a particular kind of extended participation). While this report gives national figures for many levels of participation, a pilot program will require the component involved to gather more specific indicators of current fill rates, training levels, and shortages, as well as metrics on the participation of new groups to be targeted in the likely area of impact.

Performance Monitoring

Because most of the constructs are intended to work on both supply and demand objectives, measures of effectiveness for pilot programs should be designed for the same areas described in the initial assessments. These can include both quantitative measures of reservist participation and surveys or other means of understanding the motivations, expectations, and responses of participating individuals. Measures of effectiveness should also consider differences between units where participating reservists are coming through traditional sourcing and those where a substantial number are coming via the workforce construct.

Evaluation

While there may be pressure to make a rapid determination of the pilot program's effectiveness and determine whether it justifies a broader or more permanent policy change, it will be equally important for the pilot's sponsor to conduct a thorough final evaluation. In some cases, these innovations may not have already been tried due to concerns about long-term impacts. For example, the Seasonal Worker, Seasonal Reservist model might work well for a year or two, and then the workers might decide the pace is unsustainable. Alternatively, the pilots might have unexpected benefits. (Some seasonal workers may only keep this schedule for a few years and then opt for more traditional jobs; it would be interesting to see if they also switch to a traditional reserve careers as well.) Key evaluative criteria for each pilot therefore include the following questions:

- Is the pilot meeting established aims in terms of recruitment and retention of the target population to fill identified personnel gaps?
- Does the pilot have unintended negative consequences, and if so, do such consequences outweigh the benefits experienced in terms of recruitment and retention of a broader spectrum of the U.S. workforce?
- What are the short-term and projected long-term impacts of the pilot, and to what extent do the benefits of the pilot appear to be sustainable?

In order to gauge both short-term and long-term effects of the pilot, we recommend that each pilot program be evaluated at least once annually.

CHAPTER EIGHT

Conclusions and Recommendations

As noted in Chapter One, this study aimed to assist the Office of the Assistant Secretary of Defense for Reserve Integration in exploring broad programmatic improvements to enhance the manpower available for a variety of national requirements. To do so, the research embodied in this report aimed to identify (1) those specialties that are most difficult to recruit and retain within the Total Force, (2) those aspects of the potential military workforce who could participate to a greater degree, and (3) the policies needed to connect those unlikely reservists to those unmet requirements. The purpose of this chapter is to present conclusions and recommendations in line with these study goals.

Conclusions

First, there is a large and growing segment of the U.S. population that is not currently a primary source of military manpower not because of any objective deficiencies in their intellectual capacity, physical capabilities, or patriotic spirit but because of other life choices and conditions. To change the nature of military service overall or to redefine expectations of AC service goes beyond this study. However, because RC service is defined by a movable line between military and civilian life, identifying where occupational characteristics may limit RC participation among some populations can suggest ways to move that line and expand that participation. For instance, work schedule requirements and work-life balance are shown to be associated with lower levels of participation by many occupation categories in the RC. Occupational groups with substantial levels of potential schedule or work-life conflict to participation include over 40 percent of all civilian employed workers as of 2016. Many occupations that report potential scheduling or work-life conflict with RC participation include skills that have been identified as in demand across military branches, such as pilots, clergy, cyber-related occupations, and medical scientists and practitioners.

Second, current trends suggest that Americans will continue becoming more accustomed to flexible employment options in their civilian lives and will expect and respect the option to use such options in military service. Beyond flexibility, predictability emerges as a key issue for target populations of the workforce. Therefore, options that offer more advanced notice, stability, and guarantees could appeal to such populations.

Third, this need not be a threat to military culture. Other researchers have noted that U.S. military personnel are increasingly accustomed to turning "pickup teams" into effective units, and not relying on extended predeployment training to create the necessary interpersonal relationships and processes.[1] To the degree that this remains a trend, the RCs can increase their use of creative ways to manage and train individuals, with the goal of having them gel into teams shortly before or soon after deployment, not in home station. Prior service individuals may have particularly niche skills that are useful and necessary. Moreover, interviews conducted during the course of this research also suggest that bringing people with previous military experience back into the RC workforce is gaining popularity and is increasingly recognized as desirable.

Fourth, the RCs offer a structural means to enable and facilitate experimentation with different structures and parameters for work.

---

[1]   Dennis P. Chapman, using data from Army National Guard plans in the spring of 2006, studied 13 deploying Army National Guard Brigade Combat Teams and found that of 47,336 soldiers who ultimately put "boots on ground" for them, 23 percent were fillers from the unit's home state, 2 percent came from other states, 1 percent came from the IRR, and less than 1 percent (188 soldiers) came from the Regular Army. Dennis P. Chapman, *Manning Reserve Component Units for Mobilization: Army and Air Force Practice*, Washington, D.C.: Association of the United States Army, Land Warfare Paper No. 74, September 2009, p. 9.

Each state has independent legal authority to structure Air and Army National Guard duty schedules and parameters as it chooses, provided it meets federal readiness requirements. State authority is even greater with respect to state militia or state military reserve units, over whom state governors exercise near-complete autonomy. Similarly, federal reserve components have existing legal authority to experiment with duty schedules, parameters, alternative work arrangements, recruiting processes, and other policies, provided they meet their respective services' operational requirements. The RCs may, in fact, be more able to experiment with alternative manpower policies than their AC counterparts, based on these flexibilities and the broader diversity of geographic, operational, and structural features already inherent in the reserve components.

Fifth, civilian employers have developed a range of alternative work models that enhance their ability to attract and retain talent in a competitive labor market. These include models that reduce the degree to which time and location requirements govern the employment relationship by incorporating flexibilities in terms of when and where work is performed. Younger workers in particular may value these flexibilities and seek out employers that offer them, while flexible work arrangements may also help employers to retain baby boomers reaching traditional retirement age and seeking to ramp down their workload.

Sixth, the private sector is at the forefront of developing and implementing innovative workforce models, in part because private-sector employers face fewer regulatory and budgetary restrictions on how they structure their workforces; however, nondefense federal agencies and state and local governments also utilize flexible workforce models. At the federal level, in addition to alternative work schedules (e.g., flexible or compressed work schedules) and telework, a range of part-time, temporary, and seasonal work arrangements exist across the agencies. State and local governments have long incorporated part-time, on-call, and seasonal workers in their workforces (e.g., education and recreation workers), and many state and local governments are building additional flexibilities into their full-time work offerings as well to help attract talent.

Seventh, advances in technology contribute to the development of innovative workforce models. Technology can enhance employers' ability to incorporate flexible scheduling practices, for example, by deploying scheduling analytics to identify their workforce needs or by using online and mobile technologies to allow workers to express scheduling preferences and execute shift swaps. Moreover, technology is a major driver of telework and remote work options that can broaden the pool of talent from which employers can draw. Yet it is critical to be aware that a major challenge to proposed workforce constructs involving telework or remote connections are the availability, functionality, and security of appropriate technology and systems to enable such modes of working.

Recommendations

The purpose of this study was to stimulate discussion on alternative approaches to manning the RCs, and the breadth of this initial effort meant that none of the possible workforce constructs were presented as a "ready to wear" solution. Any of them would need to be fully staffed and coordinated before they could be implemented, even as a pilot program. However, we feel that any of them *could* be implemented, if one or more services agreed that it would be worth applying to solve their particular manpower shortage, and recommend that OSD actively encourage such service-level efforts.

We have provisionally placed the pilot projects in three bins, in order of increasing barriers to experimentation in the near future:

1. *Getting comfortable with technology.* The Telereserve and Reserves on Demand programs would probably grow incrementally as new tools from the civilian sector become available and as military managers see advantages to adopting them.
2. *Waiting for an advocate.* The Warrant Officer–Deacon, Wounded Warrior, and Seasonal Worker, Seasonal Reservist concepts seem to have few institutional barriers. These workforce constructs could be implemented by a service with leadership support.

3. *Changing culture, policy, and systems.* No Passport Required, Job Sharing, Part-Time Plus, and Sponsored Reserves share the feature of changing more fundamental terms of the RC experience. Creating even a workforce construct for one of these concepts would require a coalition of advocates to develop the personnel systems to manage a new category of reservists, change policies and statutes, and gain buy-in from both the AC and RCs. The programs' feasibility would depend on whether such changes added capacity to the Total Force and were viewed as worth the costs and effort.

Unless a senior DoD leader is ready to become a long-term advocate for one of the constructs in the third category, the most effective route to change would likely be one or more of the constructs in the second category, which could generate some momentum for broader changes.

Beyond recommending the adoption of particular workforce constructs, we recommend the following actions to continue developing the knowledge base on RC workforce challenges and potential solutions to adapt RC service to the current market for RC manpower:

First, OSD should continue to assess its access to required talent—now and in the future—as well as the extent to which current manpower policies enhance or reduce the propensity and ability of Americans to serve.

Second, OSD and the services should continue to assess the extent to which their workforce practices converge or diverge from common practices in the civilian workforce. There may be opportunities for DoD to learn from best practices in the civilian sector while reducing churn and turbulence for service members and dependents by harmonizing DoD HR practices with those common in the civilian workforce. In doing so, the RCs should carefully monitor trends in the civilian labor market regarding alternative work arrangements and worker preferences for time, location, and other types of flexibility. Devising and implementing new forms of RC service that draw on innovative workforce models developed by private-sector employers, nondefense federal agencies, and state and local governments will be especially important if the economy continues to expand and the RCs face significant competition for talent from a healthy civilian labor market.

Third, the services should authorize their RCs to experiment with alternative work structures where a demonstrable need exists and where the alternative work structure appears likely to meet that need. For example, if the services identify a problem with cyberoperator retention and believe that alternative work parameters would help alleviate the problem, the appropriate components should be encouraged and authorized to test different approaches (such as those outlined in this report). To this end, the RCs should explore new service options that both reflect and complement developments in the civilian workforce. For example, models could include time and location flexibility beyond what is currently available, in line with flexible work arrangements developed outside the military. These same models could allow enhanced participation by individuals with flexible or intermittent civilian work schedules that do not adhere to the traditional concept of the 40-hour workweek with weekends free.

Fourth, the RCs should regularly consider how technological innovation can promote greater innovation in when and where individuals perform their service and the extent to which they need to be present at the same time in the same place to train successfully. To that end, each service should explore how it could expand existing telework programs under the broad authority of DoDI 1035.1 to allow reservists to perform a broader set of tasks remotely. The RCs should also consider how technology could facilitate the matching of RC members with discrete tasks that need to be completed, whether remotely or in person.

Finally, OSD and the services should continue to support such efforts as duty-status reform, which will add more flexibility and simplicity to the system and mirror advances in reserve force management adopted by allied countries, such as Australia.

Identifying Shortfalls in Specialties Across the Services

To better understand the needs of the services, we reviewed recent congressional testimony before the House and Senate Armed Services Committees, government reports, previous studies or reports, and government-commissioned research to identify current and anticipated manpower shortfalls for each service, specialties that are difficult to recruit and/or retain, and general trends in military personnel requirements. The intent of this review was not to prepare a comprehensive list or to determine precise requirements. Rather, we sought to gain an overall impression of the types of unmet needs the services are facing and help target potential areas that would benefit from alternative work arrangements.

The review produced a wide variety of information with varying levels of fidelity, timeliness, and granularity. It was challenging to find direct comparisons across the Army, Navy, Air Force, and Marine Corps for approximately the same time. To provide roughly comparable information in this report, we have produced a table with information from congressional testimony, a second table with published lists of enlisted specialties eligible for enlistment bonuses or retention/reenlistment bonuses in 2018, and findings from several recent GAO reports.

Table A.1 attempts to provide roughly comparable information, although different individuals providing testimony would provide varying levels of specificity, contributing to some unevenness. It includes information provided in written testimony and provided

Table A.1
Shortages Reported in Congressional Testimony

Hearing	Army	Navy	Air Force	Marine Corps	Across Services
April 2018, House Armed Services Committee, Military Personnel Subcommittee[a]	Aviation Cyber	Nuclear field Nuclear propulsion Special warfare Advanced electronics Aviation Cyber Linguists Mechanical and structural Health care professions Aviation maintenance IT specialists Submarine officers Surface warfare officers Strike fighter Electronic attack Helicopter mine countermeasure Maritime patrol reserves Fleet logistics reserves	Pilots (short 2,000, including 1,300 fighter pilots) Linguists Special operations aircraft maintenance Nuclear medicine Combat systems officers Various medical specialties	Cyber High-tech occupations Aviation Aviation maintenance Collateral duty inspectors Cybersecurity technicians Special operators Counterintelligence specialists	Aviation Cyber

Table A.1—Continued

Hearing	Army	Navy	Air Force	Marine Corps	Across Services
May 2017, House Armed Services Committee, Military Personnel Subcommittee[b]	Cyber operators	Nuclear-trained enlisted personnel	Cyber operators (defense)	Cyber	STEM
		Linguists	Battlefield airmen	Human intelligence	
		Cryptology specialists	Intelligence experts	Counterintelligence	
		Cyber operations (offense and defense)	Explosive ordnance disposal	High-demand/low-density MOSs	
		Information warfare	Nuclear enterprise specialists	Highly technical MOSs	
		Advanced electronics fields	Pilots	Cybersecurity technicians	
		Special warfare	Maintenance (enlisted and officer)		
		Unrestricted line, restricted line, and staff corps officers	Special operators		
		Aviation	Contracting		
		Surface warfare officers	Selected health professions		
		Submarine department head			
		Fighter pilots			
		Chaplains			

[a] Data from Robert P. Burke, Chief of Naval Personnel, U.S. Navy, testimony before the U.S. House Armed Services Committee, Subcommittee on Military Personnel, April 13, 2018.

[b] Data from Robert P. Burke, Chief of Naval Personnel, U.S. Navy testimony before the U.S. House Armed Services Committee, Subcommittee on Military Personnel, May 17, 2017.

orally during the hearing from members and witnesses in terms as close to the transcript or text as possible. The rightmost column captures items that members or witnesses identified as being shortfalls across all the services.

Exploration of those fields for which each service is offering an enlistment or retention bonus serves as another means of identifying shortage specialties. This is a fairly analytically rigorous method of generating a list of specific MOSs, as the services typically base the bonus amount on historical trends, alternative ways to meet the demand, and the operational impact of a shortage in each specialty. Table A.2 shows these MOSs for each service, as of mid-2018. Note, however, that MOS shortages in the AC and RC may not necessarily align. (Some MOSs do not exist in all components, or the same MOS may be easy to fill in one component and hard in another.) MOSs listed in Table A.2 are largely AC specialties.

Government Accountability Office–Documented Shortages Across the Services

GAO has highlighted shortages in several key occupations across the services. First, there is a pilot shortage across the services, particularly among fighter pilots. Two 2018 GAO studies found increasing shortages of fighter pilots across the Air Force, Navy, and Marine Corps.[1] One of the studies found that the Air Force had the largest gap, with only 73 percent of authorizations filled in its AC billets in FY 2017. This was down from 95 percent in FY 2006. The Navy and Marine Corps showed similar trends. The Navy had filled 88 percent of authorizations in FY 2013 but only 74 percent in FY 2017. Marine Corps fill rates declined from 94 percent in FY 2006 to 76 percent in FY 2017.[2] In addition to identifying the same gaps for fighter pilots across the ser-

---

[1] GAO, *Military Personnel: DOD Needs to Reevaluate Fighter Pilot Workforce Requirements*, Washington, D.C., GAO-18-113, April 11, 2018b; GAO, *Military Personnel: Collecting Additional Data Could Enhance Pilot Retention Efforts*, Washington, D.C., GAO-18-439, June 2018c.

[2] GAO, 2018b.

Table A.2
Enlistment or Selective Retention Bonus Specialties, by Service

Service	Service Code	Specialty
Army	12N	Horizontal construction engineer
	13B	Cannon crew member
	13M	Multiple-Launch Rocket System crew member
	13R	Field artillery Firefinder radar operator
	14E	Patriot fire control enhanced operator/maintainer
	14G	Air defense battle management system operator
	14H	Air defense enhanced warning system operator
	14P	Air and missile defense crew member
	14T	Patriot launching station enhanced operator/maintainer
	25N	Nodal network systems operator/maintainer
	25P	Microwave systems operator/maintainer
	25Q	Multichannel transmission systems operator/maintainer
	25S	Satellite communication systems operator/maintainer
	25U	Signal support systems specialist
	25G	Geospatial intelligence imagery analyst
	35M	Human intelligence collector
	35N	Signals intelligence analyst
	35Q	Cryptologic cyberspace intelligence collector/analyst
	35P	Cryptologic linguist
	35T	Military intelligence systems maintainer/integrator
	42R	Musician
	68K	Medical laboratory specialist
	68R	Veterinary food inspection specialist
	68S	Preventative medicine specialist

Table A.2—Continued

Service	Service Code	Specialty
	74D	CBRN specialist
	88M	Motor transport operator
	89D	EOD specialist
	91J	Quartermaster and chemical equipment repairer
	91M	Bradley Fighting Vehicle system maintainer
	92A	Automated logistical specialist
	92F	Petroleum supply specialist
	92G	Culinary specialist
	92R	Parachute rigger
	94D	Air traffic control equipment repairer
	94E	Radio and communications security repairer
	94F	Computer/detection systems repairer
	94H	Test measurement and diagnostic equipment maintenance support specialist
	94M	Radar repairer
	94T	Short-range air defense system repairer
	94S	Patriot system repairer
	94Y	Integrated Family of Test Equipment operator/maintainer
Navy	ABE, ABF, ABH	Aviation boatswain's mate (launching and recovery equipment, fuels, aircraft handling)
	IT	Information systems technician
	AM	Aviation structural mechanic
	IS	Intelligence specialist
	AO	Aviation ordnanceman
	ND	Navy diver
	BM	Boatswain's mate

Table A.2—Continued

Service	Service Code	Specialty
	EM, ET, MM	Nuclear program
	CTI, CTM	Cryptologic technician interpretive, maintenance
	OS	Operations specialist
	CTN, CTR	Cryptologic technician network, collection
	CB	SeaBees
	EOD	Explosive ordnance disposal
	SO	Special operations
	FC	Fire controlman
	SB	Special operations boat
	FT	Fire control technician
	STG	Sonar technician surface
	HM	Hospital corpsman
	STS	Sonar technician submarine
Air Force	1A0X1	In-flight refueling
	1A1X1	Flight engineer
	1A6X1	Flight attendant
	1A8X1	Airborne cryptologic language analyst (Arabic, Chinese, Korean, Russian, Spanish, Hebrew)
	1A8X2	Airborne intelligence, surveillance, and reconnaissance operator
	1B4X1	Cyber warfare operations
	1C2X1	Combat control
	1C3X1	Command and control operations
	1C4X1	Tactical air control party
		All-source intelligence analyst

Table A.2—Continued

Service	Service Code	Specialty
	1N0X1	Geospatial intelligence/targeteer
	1N3X1	Cryptologic language analyst (Arabic, Chinese, Korean, Russian)
	1N4X1A	Fusion analyst/digital network analyst
	1N7X1	Human intelligence specialist
	1S0X1	Safety
	1T0X1	Survival, evasion, resistance, and escape
	1T2X1	Pararescue
	1W0X2	Special operations weather
	2A2X1A	Special operations forces/personnel recovery integrated communications/navigation/mission systems
	2A373, 2A377, 2A3X3, 2A3X7	Tactical and fifth generation aircraft maintenance (A-10, U-2, F-15, F-16, F-22, F-35)
	2A378, 2A3X7B	Remotely piloted aircraft maintenance (MQ-1, MQ-9)
	2A572, 2A5X2D	Helicopter/tilt-rotor aircraft maintenance (CV-22)
	2A5X1A	Airlift/special mission aircraft maintenance (C-20, C-21, C-22, C-37, C-40, E-4, VC-25)
	2A554B, 2A554C, 2A554E	Refuel/bomber aircraft maintenance (KC-10, KC-46, C-1)
	2A7X5	Low observable aircraft structural maintenance
	2A374, 2A375, 2A3X4, 2A3X5	Fighter and advanced fighter aircraft integrated avionics (A-10/U-2, F-15, F-16, F-22, F-35, MQ-1, MQ-9, RQ-4)
	2A871J, 2A8X1F, 2A8X1G, 2A8X1H	Mobility air forces integrated communications/counter/navigation systems (tanker communications/counter/navigation system, KC-10, KC-135, KC-46)

Table A.2—Continued

Service	Service Code	Specialty
	2A8X2F, 2A852G, 2A8X2H	Mobility air forces integrated instrument and flight control systems (KC-10, KC-135, KC-46)
	2A951B, 2A9X1E, 2A9X1F, 2A9X1G	Bomber/special integrated communication/navigation/ mission systems (E-V/VC-25, B-1, B-2, B-52)
	2A952A, 2A9X2B, 2A9X2D, 2A952E	Bomber/special integrated instrument and flight control systems (E-3, V-4/ VC-25, RC-135, B-1)
	2A953B, 2A9X3F	Bomber/special electronic warfare and radar surveillance integrated avionics (E-3 computer/electronic warfare, E-2)
	2W2X1	Nuclear weapons
	3D1X4	Spectrum operations
	3E8Xa	Explosive ordnance disposal
	3F0X1	Personnel
	3F2X1	Education and training
	3F3X1	Manpower
	3F4X1	Equal opportunity
	4A2X1	Biomedical equipment
	4M0X1	Aerospace and operational physiology
	4N0X1B, 4N0X1C, 4N0X1F	Aerospace medical service (neurodiagnostic medical technician) independent duty medical technician, flight and operational medical technician
	4N1X1B, 4N1X1C, 4N1X1D	Surgical service (urology, orthopedics, otolaryngology)
	4P0X1	Pharmacy
	4R0X1A, 4R0X1B, 4R0X1C	Diagnostic imaging (nuclear medicine, diagnostic medical sonography, magnetic resonance imaging)

Table A.2—Continued

Service	Service Code	Specialty
	4T0X1	Medical laboratory
	4Y0X1H	Dental hygienist
	4Y0X2	Dental laboratory
	5R0X1	Chaplain assistant
	7S0X1	Special investigations
	0211	Counterintelligence/human intelligence marine
Marine Corps	0321	Reconnaissance man
	0372	Critical skill operator
	0659	Cyber network systems chief
	0651	Cyber network operator
	0689	Cyber security technician
	2336	Explosive ordnance disposal technician
	5821	Criminal investigator agent
	7257	Air traffic controller
	0241	Imagery analysis specialist

SOURCES: U.S. Army, "Benefits," webpage, undated; U.S. Department of the Navy, Fiscal Year 2019 Budget Estimates, Justification of Estimates, February 2018; U.S. Air Force Personnel Center, "Selective Retention Bonus Listing," May 30, 2018.

[a] Service codes are MOSs for the Army, Navy, and Marine Corps and AFSCs for the Air Force.

vices, the second study identified a persistent gap in Navy surveillance and transport specialties.[3]

GAO also documented a widespread shortage of military medical professionals. A February 2018 GAO reported persistent AC and RC physician shortfalls (less than 80 percent of authorizations filled

---

[3]   GAO, 2018c.

in FYs 2011–2015) across the services: 39 components reported shortages across 19 specialties, including 11 specialties designated as critically short wartime specialties. The greatest shortage was in residency-trained aviation/aerospace medicine, with 7 components (all but the Navy Reserve) persistently filled below 80 percent of authorizations. The Navy Reserve and Army Reserve had the largest number of specialties with shortages, at 13 and 11, respectively.[4]

Feedback from the Services

We requested information from the Army, Navy, Air Force, and Marine Corps, as well as interviews with representatives from their respective policy analysis organizations. Of particular interest for our study were current and anticipated personnel shortages, especially in the RCs. The information in the following was derived from interviews with Army and Navy representatives and data provided by these services. While all four services responded to this data request to at least some extent, the Air Force and Marine Corps did not provide sufficiently extensive data to inform our assessments of shortage specialties beyond what we found in the literature and policy review.

Army

Drawing from conversations with Army representatives, Table A.3 lists shortages in the Army Reserve and Army National Guard for enlisted personnel, warrant officers, and commissioned officers.

Navy

In an interview, Navy representatives indicated that there were enlisted shortages across domains. For example, for surface operations, there were shortages of engineers, electronic technicians, and gas turbine specialists; in aviation, there were shortages of naval aircrewmen (helicopter) (AWS) and naval aircrewmen operators (AWO); and there were general shortages in special operations forces.[5] Officer shortages were

---

[4] GAO, *Military Personnel: Additional Actions Needed to Address Gaps in Military Physician Specialties*, Washington, D.C., GAO-18-77, February 2018a.

[5] Telephone interview with Navy representatives, July 13, 2018.

Table A.3
Shortages in the Army National Guard and Army Reserve

Enlisted		Warrant Officer		Commissioned Officer	
MOS	Specialty	MOS	Specialty	MOS	Specialty
88M	Motor transport operator	120A	Construction engineering technician	27A	Judge advocate general attorney
74D	Chemical operations specialist	255A	Information services technician	90A	Multifunctional logistician
88H	Cargo specialist	915A	Automotive maintenance warrant officer	88A	Transportation officer
92F	Petroleum supply specialist	920A	Property accounting technician	12A	Engineer officer
12N	Horizontal construction engineer	919A	Engineer equipment maintenance warrant officer	66H	Medical-surgical nurse
68W	Health care specialist	170A	Cyberoperations technician	38G	Military government specialist
88N	Traffic management coordinator	131A	Field artillery technician	74A	CBRN officer
91B	Wheeled vehicle mechanic	351M	Human intelligence collection technician	38A	Civil affairs officer
92W	Water treatment specialist	255N	Network management technician	35D	Military intelligence officer
31B	Military police	350F	All source intelligence technician	46A	Public affairs officer
11B	Infantrymen	921A	Airdrop systems technician	13A	Field artillery officer
13B	Cannon crew member	923A	Petroleum systems technician	91A	Ordnance officer
13J	Fire control specialist	353T	Military intelligence systems maintenance/ integration technician	11A	Infantry officer

Table A.3—continued

Enlisted		Warrant Officer		Commissioned Officer	
MOS	Specialty	MOS	Specialty	MOS	Specialty
11C	Indirect fire infantryman	948D	Electronic missile systems maintenance warrant officer	65D	Physician assistant
13F	Fire support specialist	890A	Ammunition warrant officer	88A	Transportation officer
19D	Cavalry scout	290A	Electronic warfare technician	92A	Indirect fire infantryman
25P	Microwave systems operator/maintainer	140A	Command and control systems technician	42B	HR officer (adjutant general officer)
91G	Fire control repairer	352N	Signals intelligence analysis technician	72D	Environmental science/engineering officer
12G	Quarrying specialist	914A	Allied trades warrant officer	70K	Medical logistics officer
35P	Cryptologic linguist	125D	Geospatial engineering technician	17A	Cyberoperations officer
92S	Shower/laundry and clothing repair specialist			91A	Ordnance officer
94P	Multiple-Launch Rocket System repairer			89E	EOD officer
94M	Radar repairer			36A	Financial manager
94A	Land combat electronic missile system repairer				
15Y	AH-64D armament/ electrical/avionic systems repairer				
94R	Avionic and survivability equipment repairer				

found among SEALs, medical professionals, nurses, foreign affairs officers, surface warfare officers, Medical Service Corps, dental, and aviation maintenance duty officers. The interviewees identified SEALs and medical specialties as persistent shortages over a long period. When asked what they might anticipate as future shortages, they identified special operations, medical, dental, and other health professionals. They also recognized the growing shortage of naval aviators in the AC.

Detailed Analysis of Potential Human Capital Rewards of Alternatives

The tables presented in this appendix are the unabridged versions (including all occupational groups) of those that appear in Chapter Four. The AWCS and the GSS were each used to examine potential links between work schedules, work-life balance, and occupational groups. The ACS was also used to estimate the number of individuals currently employed in each occupational category by age and sex. The AWCS, GSS, and ACS are probability sample surveys, and population weights have been applied in the analyses. For additional discussion of the tables and interpretation of their content, please refer to Chapter Four.

American Working Conditions Survey and General Social Survey Analysis

These tables are unabridged versions of those that appear in Chapter Four. The tables present the weighted proportion of respondents within each occupational category that report the job-related characteristics in each column.

The share of AWCS and GSS respondents across occupations whom we coded as having work-life conflicts varied by question, ranging from 9 percent with irregular or on-call schedules to 41 percent who often or very often felt used up at the end of the day, according to GSS data. About one-quarter of workers reported working extra hours ten or more days per month, that when they worked extra hours it was required by their employer, and that it was hard to take time off from their jobs. Lower shares reported that their job interfered with their family life (13 percent) or that it was difficult to fulfill family responsibilities several times a week because of work (11 percent). Regarding work schedules, the share rises from 9 to 17 percent when including those who typically work split or rotating schedules along with those with irregular or on-call schedules.

American Community Survey Analysis

This analysis compares the percentage of individuals reporting RC service by occupational group and overall. Percentages are presented by sex for all ages, as well as for ages under 41 in an effort to focus on those in RC-eligible ages. Occupations with lower participation in the RCs than the national average may indicate potential conflicts between occupation and RC service. Note that all differences in proportions reporting participation in the RCs versus the average across all occupations are significant at the $p < .001$ level, with the exception of women under age 41 in "Food preparation and service related" and "Health care and related technical," which are substantially different from the average across all occupations at the $p < .01$ level. In the GSS, workers in legal occupations were most likely to report feeling used up and were second most likely to report working extra hours ten or more days per month. Possibly reflecting these trends, in the ACS, lawyers, judges, magistrates, and other judicial workers reported participating in the RCs at rates of half the national average. Workers in computer and mathematical occupations were most likely to report both feeling too tired to do chores and that it was difficult to fulfill family responsibilities. The ACS data indicate that software developers, computer and information research scientists, and computer systems analysts have substantially lower RC participation rates than the national average. In the AWCS, farm workers reported the highest prevalence of seasonal work and irregular days in their typical work schedules, as well as conflict between working hours and family and social commitments and

Table B.1
Work-Life Characteristics from the American Working Conditions Survey, 2015

Occupational Group	Usually Works 6 or 7 Days per Week	Works Different Number of Days Every Week	Schedule Set by Employer with No Possibility for Changes	Predictable Seasonal Work During the Year	Unpredictable or Irregular Work During the Year	Job Involves Tight Deadlines All/Almost All the Time	Working Hours Do Not Fit Very Well or at All with Family/Social Commitments Outside of Work	Too Tired for Activities in Private Life Always/Most of the Time
All occupations	11%	30%	36%	11%	10%	34%	16%	22%
Administrative support and retail sales	20%	24%	39%	8%	9%	48%	25%	29%
Executive, administrative, and managerial	12%	30%	36%	13%	7%	28%	14%	21%
Farm operators and managers	13%	55%	34%	24%	2%	14%	41%	33%
Financial sales and related	2%	27%	20%	3%	6%	30%	4%	6%
Food preparation and service	45%	26%	0%	0%	0%	29%	0%	0%

Table B.1—Continued

Occupational Group	Usually Works 6 or 7 Days per Week	Works Different Number of Days Every Week	Schedule Set by Employer with No Possibility for Changes	Predictable Seasonal Work During the Year	Unpredictable or Irregular Work During the Year	Job Involves Tight Deadlines All/Almost All the Time	Working Hours Do Not Fit Very Well or at All with Family/Social Commitments Outside of Work	Too Tired for Activities in Private Life Always/ Most of the Time
Management-related	9%	30%	29%	9%	10%	35%	16%	18%
Professional specialty	18%	14%	49%	2%	6%	23%	13%	24%
Technicians and related support	12%	26%	3%	0%	4%	38%	0%	14%
Transportation, construction, mechanics, mining, and agriculture	13%	50%	30%	22%	2%	13%	37%	30%

SOURCE: American Working Conditions Survey, American Life Panel data.

Table B.2
Work-Life Characteristics from the General Social Survey Quality of Working Life Module

Occupational Group	Irregular Shift/On-Call Schedule	Irregular Shift/On-Call, Split Shifts, or Rotating Shifts	Works Extra Hours 10 or More Days per Month	Extra Hours Are Required by Employer	Somewhat Hard or Very Hard to Take Time Off	Demands of Job Often Interfere with Family Life	Often Feels "Used Up" at the End of the Day	Comes Home Too Tired for Chores	Difficult to Fulfill Family Responsibilities
All occupations	9%	17%	23%	27%	27%	13%	41%	28%	11%
Architecture and engineering	6%	10%	37%	25%	14%	11%	29%	28%	0%
Arts, design, entertainment, sports, and media	21%	30%	22%	27%	18%	10%	33%	25%	9%
Building and grounds cleaning and maintenance	6%	10%	12%	22%	20%	9%	32%	6%	0%
Business and financial operations	11%	15%	27%	25%	19%	11%	37%	33%	4%
Community and social service	14%	20%	26%	24%	23%	17%	44%	33%	0%
Computer and mathematical	6%	9%	29%	22%	14%	8%	34%	56%	31%

Table B.2—Continued

Occupational Group	Irregular Shift/ On-Call Schedule	Irregular Shift/On-Call, Split Shifts, or Rotating Shifts	Works Extra Hours 10 or More Days per Month	Extra Hours Are Required by Employer	Somewhat Hard or Very Hard to Take Time Off	Demands of Job Often Interfere with Family Life	Often Feels "Used Up" at the End of the Day	Comes Home Too Tired for Chores	Difficult to Fulfill Family Responsibilities
Construction and extraction	10%	12%	27%	34%	25%	13%	43%	19%	16%
Education, training, and library	4%	6%	37%	18%	38%	11%	45%	34%	19%
Farming, fishing, and forestry	4%	9%	17%	41%	24%	11%	39%	29%	0%
Food preparation and serving related	10%	32%	12%	25%	33%	11%	48%	28%	19%
Health care and related technical	9%	20%	20%	27%	41%	14%	45%	43%	23%
Health care support	11%	21%	11%	13%	30%	7%	39%	40%	8%
Installation, maintenance, and repair	8%	11%	27%	36%	22%	15%	39%	18%	10%
Legal	8%	8%	39%	24%	19%	15%	54%	0%	0%
Life, physical, and social sciences	11%	18%	22%	20%	18%	9%	42%	25%	13%

Table B.2—Continued

Occupational Group	Irregular Shift/ On-Call Schedule	Irregular Shift/On-Call, Split Shifts, or Rotating Shifts	Works Extra Hours 10 or More Days per Month	Extra Hours Are Required by Employer	Somewhat Hard or Very Hard to Take Time Off	Demands of Job Often Interfere with Family Life	Often Feels "Used Up" at the End of the Day	Comes Home Too Tired for Chores	Difficult to Fulfill Family Responsibilities
Management	10%	15%	40%	26%	24%	17%	40%	29%	11%
Office and administrative support	4%	9%	19%	24%	23%	8%	42%	28%	10%
Personal care and service	15%	21%	11%	18%	32%	10%	36%	30%	2%
Production	4%	13%	22%	40%	33%	12%	44%	25%	4%
Protective service	10%	40%	14%	44%	33%	26%	41%	34%	16%
Sales and related	14%	27%	20%	21%	27%	14%	39%	28%	12%
Transportation and material moving	16%	26%	22%	37%	39%	15%	44%	23%	14%

SOURCE: NORC, University of Chicago, General Social Survey data.

being too tired for activities outside of work. All the agricultural occupations reported participating in the RCs at substantially lower levels than the national average, according to the ACS data.

Table B.3
Reserve Participation by Occupation, American Community Survey, 2011–2015

Occupational Group	Ever in the Reserves (men)	Ever in the Reserves (women)	Ever in the Reserves (men age <41)	Unweighted Sample Size (men age <41)	Ever in the Reserves (women age <41)	Unweighted Sample Size (women age <41)
All occupations	1.5%	0.5%	0.8%	396,493	0.5%	378,180
Architecture and engineering	1.6%	0.4%	0.7%	8,622	0.3%	2,089
Arts, design, entertainment, sports, and media	1.1%	0.3%	0.4%	8,264	0.3%	7,908
Building and grounds cleaning and maintenance	1.1%	0.4%	0.6%	19,811	0.3%	10,915
Business and financial operations	1.9%	0.4%	0.7%	11,963	0.3%	16,465
Community and social service	1.6%	0.5%	0.9%	3,598	0.3%	7,585
Computer and mathematical	1.1%	0.4%	0.8%	12,609	0.2%	785
Construction and extraction	1.1%	0.3%	0.6%	27,978	0.5%	4,227

Table B.3—Continued

Occupational Group	Ever in the Reserves (men)	Ever in the Reserves (women)	Ever in the Reserves (men age <41)	Unweighted Sample Size (men age <41)	Ever in the Reserves (women age <41)	Unweighted Sample Size (women age <41)
Education, training, and library	1.5%	0.4%	0.7%	12,224	0.3%	33,329
Farming, fishing, and forestry	1.0%	0.1%	0.4%	6,811	0.2%	2,110
Food preparation and serving related	0.8%	0.4%	0.6%	31,895	0.4%	37,096
Health care and related technical	1.5%	0.6%	1.0%	8,000	0.6%	26,898
Health care support	1.4%	0.5%	1.2%	2,821	0.4%	18,466
Installation, maintenance, and repair	1.5%	0.6%	1.0%	20,400	0.7%	888
Legal	2.3%	0.5%	0.4%	2,755	0.5%	4,077
Life, physical, and social sciences	1.6%	0.4%	1.0%	3,588	0.3%	3,540
Management	2.0%	0.5%	1.0%	26,198	0.5%	20,343
Office and administrative support	1.4%	0.4%	0.8%	32,702	0.4%	69,935

Table B.3—Continued

Occupational Group	Ever in the Reserves (men)	Ever in the Reserves (women)	Ever in the Reserves (men age <41)	Unweighted Sample Size (men age <41)	Ever in the Reserves (women age <41)	Unweighted Sample Size (women age <41)
Personal care and service	1.3%	0.4%	0.8%	8,701	0.4%	26,173
Production	1.4%	0.5%	0.8%	30,218	0.4%	11,413
Protective service	2.4%	1.0%	2.2%	13,443	1.1%	4,738
Sales and related	1.8%	0.4%	0.9%	39,162	0.4%	53,227
Transportation and material moving	1.6%	0.4%	0.9%	37,840	0.3%	7,491

SOURCE: U.S. Census Bureau, American Community Survey data.

Background and Case Studies of Nonstandard Work Arrangements

Categorizing Nonstandard Work Arrangements

GAO has previously attempted to characterize the share of the workforce involved in various types of alternative or contingent work arrangements, drawing on the findings from several past surveys including the CWS, Current Population Survey, GSS, and Survey of Income and Program Participation. GAO found in 2015 that "the size of the contingent workforce can range from less than five percent to more than one-third of the total employed labor force, depending on the definition of contingent work and the data source."[1] GAO identifies a "core contingent workforce" of agency temps, direct-hire temps, on-call workers, and day laborers and finds that 7.9 percent of workers fell into this "core contingent workforce" group as of 2010.

Other sources that describe and characterize nonstandard work arrangements range from studies on how the "standard employment model" began to unravel in the 1970s[2] to the former head of the U.S.

---

[1]   GAO, "Contingent Workforce: Size, Characteristics, Earnings, and Benefits," memorandum to Sen. Patty Murray, ranking member, U.S. Senate Committee on Health, Education, Labor, and Pensions, and Sen. Kirsten Gillibrand, GAO-15-168R, April 20, 2015.

[2]   Defining "standard work arrangements" as those "in which it was generally expected that work was done full-time, would continue indefinitely, and was performed at the employer's place of business under the employer's direction." Arne L. Kalleberg, "Nonstandard Employment Relations: Part-Time, Temporary, and Contract Work," *Annual Review of Sociology*, Vol. 26, 2000.

Department of Labor's Wage and Hour Division David Weil's recent work on what he calls the "fissured workplace" in which "the employment relationship has been broken into pieces" through the use of contracting, temp agencies, and other models.[3] To Weil, "technology and software algorithms enable companies to further outsource significant proportions of the work," but fissuring long predated the development of these new technologies. A 2017 National Academies' report noted that "nontraditional types of employment—other than the 40-hour-per-week job at a single company offering health and retirement benefits—appear to be increasing" and that "while nontraditional work as independent contractors and temporary agency employees has been growing for decades, IT advances now make it easier to access such employment opportunities, and in some cases to perform work remotely over the Internet."[4]

In reviewing nonstandard work arrangements in use in the private sector and other U.S. public organizations, we came across several categorization schemes for such arrangements, confirming, as one recent paper notes, that "there is no single taxonomy to uniformly describe standard and nonstandard work arrangements."[5] As mentioned earlier, one commonly used delineation of workers in alternative work arrangements derives from the BLS CWS and includes independent contractors, on-call workers, temporary help agency workers, and workers provided by contract firms. As described by BLS in the 1990s when its CWS was developed, these arrangements were intended to capture "individuals whose employment is arranged through an employment

---

[3]   David Weil and Tanya Goldman, "Labor Standards, the Fissured Workplace, and the On-Demand Economy," *Perspectives on Work*, 2016.

[4]   National Academies of Sciences, Engineering, and Medicine, *Information Technology and the U.S. Workforce: Where Are We and Where Do We Go from Here?* Washington, D.C.: The National Academies Press, 2017.

[5]   John Howard, "Nonstandard Work Arrangements and Worker Health and Safety," *American Journal of Industrial Medicine*, Vol. 60, No. 1, January 2017.

intermediary such as a temporary help firm, or individuals whose place, time, and quantity of work are potentially unpredictable."[6]

Other researchers have used different methods. Susan Houseman, for example, approaches the topic from the perspective of identifying "flexible staffing arrangements" used by employers and includes short-term hires and regular part-time workers employed directly by the organization (i.e., not provided by a contractor or temporary services firm), in addition to the BLS CWS categories outlined above.[7] Sharon Mastracci and James Thompson focus their investigation on alternative work arrangements in the federal government and distinguish between "core" and "ring" jobs, with the latter category encompassing "comparatively unstable work situations," such as temporary, contract, on-call, and part-time positions.[8] The International Labour Organization defines a "non-standard form of employment" as "work that falls outside the scope of a standard employment relationship, which itself is understood as being work that is full-time, indefinite employment in a subordinate employment relationship." It places into this bucket temporary employment, temp agency and other contract work, ambiguous employment relationships, and part-time work.[9]

More systematic categorization schemes include an older rubric developed by Jeffrey Pfeffer and James Baron and a more recent one developed by Peter Cappelli and J. R. Keller. Pfeffer and Baron outline three characteristics that make a work arrangement nonstandard: limited temporal attachment, limited physical attachment, and lim-

---

[6] Anne E. Polivka, "Contingent and Alternative Work Arrangements, Defined," *Monthly Labor Review*, BLS, October 1996.

[7] Susan N. Houseman, "Why Employers Use Flexible Staffing Arrangements: Evidence from an Establishment Survey," *Industrial and Labor Relations Review*, Vol. 55, No. 1, October 2001.

[8] Mastracci and Thompson, 2009.

[9] International Labour Office, *Non-Standard Forms of Employment: Report for Discussion at the Meeting of Experts on Non-Standard Forms of Employment*, Geneva, Switzerland, February 2015.

ited administrative attachment.[10] These categories would include, for example, short-term workers (limited temporal), remote workers and telecommuters (limited physical attachment), and contractors (limited administrative attachment).

Cappelli and Keller note that "practice has eroded the usefulness of [Pfeffer and Baron's] taxonomy" and seek to further break out types of work arrangements, offering a more expansive taxonomy that differentiates among possible arrangements:[11]

- work control, which is the ability of employers to control the content of the work
- legal and personnel management responsibilities, which are all the administrative, regulatory, and legal requirements associated with workers
- involved parties, such as employer and employee or client, agency, and worker.

Figure C.1 presents Cappelli and Keller's taxonomy. Work arrangements are organized by governing legal construct—those governed by labor law and those governed by contract law. In employee-employer relationships, the employer controls both the work content (what is done) and work performance (how it is done), whereas in contract work, the client controls the work content but not work performance. In an employee-employer relationship, the relatively high transaction costs of recruiting, hiring, compensating, training, and administration suggest that these relationships will be longer term and, in effect, act as an open-ended contract.

Cappelli and Keller's taxonomy describes relationships among the actors in a range of alternative work arrangements. In addition to

---

[10] Jeffrey Pfeffer and James N. Baron, "Taking the Workers Back Out: Recent Trends in the Structuring of Employment," in Barry M. Staw and Larry L. Cummings, eds., *Research in Organizational Behavior*, Vol. 10, Greenwich, Conn.: JAI Press, 1988, cited in George and Chattopadhyay, 2017.

[11] Peter Cappelli and J. R. Keller, "Classifying Work in the New Economy," *Academy of Management Review*, Vol. 38, No. 4, 2013.

Figure C.1
Taxonomy of Employment Relationships

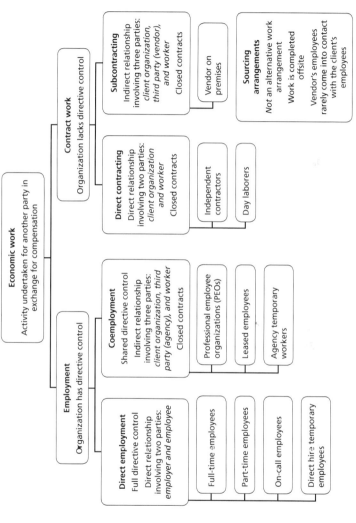

SOURCE: Peter Cappelli and J. R. Keller, "Classifying Work in the New Economy," *Academy of Management Review*, Vol. 38, No. 4, 2013.

the relationships described here, flexibility can be provided in terms of where and when the work is performed. Our review of alternative work arrangements was not limited to flexible arrangements in the context of the employer-employee relationship nor to its expected duration. Rather, we expanded our search to consider a broader set of flexibilities that work arrangements can reflect, including those that may make a job more attractive to potential employees, promote a more sustainable work-life balance, or allow for participation of parents and other care-givers in the workforce.

We considered several approaches to categorizing these flexible practices. A 2014 report by the Council of Economic Advisers described flexible workplace arrangements as follows:

> Flexible workplace arrangements refer to a broad set of firm practices that touch on when one works, where one works, or how much one works (including time off after childbirth or other life events). They include a variety of arrangements such as job sharing, phased retirement of older workers, telecommuting, and schedule shifting or flexible schedules. Workplace flexibility can be short-term, such as allowing workers to shift their work-day to end an hour earlier than usual to take a parent to a doc-tor's appointment. They can also be long-term, like allowing a reduced schedule of four and a half days per week so that a father can participate in therapy for his autistic son.[12]

SHRM has developed a rubric of types of workplace flexibilities, shown in Table C.1. The SHRM rubric also includes a discussion of the benefits and challenges of these approaches, which we summarize in the next section.

The SHRM rubric overlaps with the categories included in an older BLS review of flexible work arrangements, which identified the following types of arrangements that included but were not limited to the alternative work arrangements in the CWS: part-time work; job sharing; on-call, temporary, contract, and seasonal work; telecommut-

---

[12] Council of Economic Advisers, *Work-Life Balance and the Economics of Workplace Flexibility*, June 2014.

Table C.1
Types of Workplace Flexibility

Type of Flexibility	Type of Policy/Practice
Work schedule	• Flex time • Compressed workweek • Flex-shift work/workday schedule • Self-scheduled breaks • Part-year/seasonal work • Weekend/evening/night work
Location	• Telework, home-based work • Remove work • Hoteling
Amount of work	• Job sharing • Reduced load or customized work/part-time work
Continuity of work	• Long-term breaks/sabbaticals, career flexibility • FMLA • Compensatory time off

SOURCE: Ellen Ernst Kossek, Leslie B. Hammer, Rebecca J. Thompson, and Lisa Buxbaum Burke, *Leveraging Workplace Flexibility for Engagement and Productivity*, Society for Human Resource Management Foundation, 2014.

ing, teleworking, and flex placing; shift work; and flexible scheduling, compressed workweeks, and compensatory time off.[13] It also dovetails with the categorization scheme proposed by Georgetown University's Workplace Flexibility 2010 project, which considers flexible work arrangements to be those with flexibility in terms of the scheduling of hours, the amount of hours worked, or the place of work.[14] The Organisation for Economic Co-operation and Development also has outlined "family-friendly workplace arrangements" and includes in this category "extra-statutory leave" from work arrangements (leave other than that required by law); employer-provided childcare, "out-of-school-hours"

---

[13] Torpey, 2017.

[14] Workplace Flexibility 2010 at Georgetown University Law Center, "Flexible Work Arrangements: A Definition and Examples," undated.

care, and elderly care supports; and flexible working time arrangements, such as teleworking and working nonstandard hours.[15]

While none of these sources should be considered authoritative or comprehensive, especially in light of ongoing innovations in the workplace, including new technologies, collectively they give a sense for the types of nonstandard work arrangements and types of workplace flexibilities that exist and that have been analyzed by researchers.

Potential Benefits and Drawbacks of Nonstandard Work Arrangements

Work arrangements other than full-time, permanent work at a fixed employer location can have both benefits and drawbacks—for both the employer and the individual performing the work. While the pendulum can swing depending on the type of flexibility in question, perhaps the core trade-off is between flexibility (for employers to construct their workforce such that it aligns with the ebb and flow of work or for individuals to build work schedules that mesh with their nonwork responsibilities) and stability (for employers seeking to cultivate and retain talent or for workers looking for reliable, consistent work in a country in which health, retirement, and other benefits continue to be offered through an employer-provided model).

A recent paper published by SHRM and the Society for Industrial and Organizational Psychology (SIOP) outlines the pros and cons for employers and workers of nonstandard work arrangements in general.[16] For employers, benefits can include cost savings (either by offering lower wages or cutting down on facilities costs by utilizing off-site workers), increased flexibility to align the supply of labor with the demands of the organization's work flow, and enhanced ability to use technology to coordinate across place and time. These potential benefits echo those discussed in Houseman (2001), which found that among the reasons

---

[15] Organisation for Economic Co-operation and Development, "LMF2.4. Family-Friendly Workplace Policies," OECD Family Database, December 8, 2016.

[16] George and Chattopadhyay, 2017.

employers utilized flexible staffing arrangements were to "adjust for workload fluctuations and staff absences," to screen workers for permanent positions, and to trim spending on employer-provided benefits.[17] Earlier work by Houseman (1999) further noted that "sometimes employers use flexible staffing arrangements to access workers with special skills" and suggested that increasingly complex technology may motivate firms to increase their use of flexible arrangements to draw in talent.[18] The International Labour Organization finds a similar set of motivations for using nonstandard workers, including an ability to have numerical flexibility (i.e., increase or decrease staff in line with changing demand patterns), functional flexibility (i.e., hire workers with skills not available in-house for short-term needs), to achieve cost savings, and to take advantage of technologies that allow the organization to put together teams in disparate locations.[19] One recent survey of hiring managers found that a talent shortage is the leading reason for employers adopting flexible workforce models, including use of temporary, agency, and freelance workers.[20]

Per the SHRM/SIOP paper, potential drawbacks for employers of nonstandard work arrangements include leading to more turnover and a need to continually onboard and train workers, increasing "coordination and integration costs," and using such work arrangements may be a sign of a "lack of commitment to the workforce."[21] Other sources describe managerial challenges that can arise when nonstandard work arrangements are utilized, discuss best practices for how organizations

---

[17] Houseman, 2001.

[18] Susan N. Houseman, *Flexible Staffing Arrangements: A Report on Temporary Help, On-Call, Direct-Hire Temporary, Leased, Contract Company, and Independent Contractor Employment in the United States*, August 1999.

[19] International Labour Office, 2015.

[20] Inavero, 2018.

[21] George and Chattopadhyay, 2017.

can most effectively utilize nonstandard workers, and identify the types of tasks best suited for flexible staffing arrangements.[22]

Workers can face a similar trade-off, with potential benefits including being able to choose when and where to work so as to enhance work-life balance and potential drawbacks in terms of feeling less connected to the organization and having the "onus for skill development" on the individual rather than the organization.[23] Flexible, nonstandard workers may also receive lower pay and fewer benefits than regular, full-time employees, though this can vary significantly, for example, with high-skilled independent contractors commanding higher wages compared with other on-call or temporary workers.[24] As the International Labour Organization writes, "Non-standard employment may therefore contribute to improved employment outcomes and a better work-life balance, provided that the working conditions are decent and that it is the worker's choice to engage in this type of employment."[25]

Importantly, the implications of flexible work arrangements for workers can vary by type of flexibility and *who controls the flexibility*. For example, when employers use "just-in-time" scheduling and other practices that afford them flexibility to manage their labor force but that offer their employees limited control over their schedules, this unpredictability can be detrimental to workers.[26] At the other end of the spectrum, a flexible work arrangement that permits employees to determine their own hours (and/or work location) as long as tasks are accomplished would constitute a benefit for workers. Indeed, workplace flexibility programs can help employers to attract and retain employees, from millennials who may be seeking a less formal, hierarchical workplace to midcareer workers looking to balance work and

---

[22] For a review of relevant literature, see Elizabeth George and Prithviraj Chattopadhyay, *Non-Standard Work and Workers: Organizational Implications*, Geneva, Switzerland: International Labour Office, Conditions of Work and Employment Series No. 61, October 14, 2015.

[23] George and Chattopadhyay, 2017.

[24] Mastracci and Thompson, 2009.

[25] International Labour Office, 2015.

[26] Golden, 2015.

home responsibilities to older workers who want to stay attached to the labor force.[27] In some cases, workers may be willing to trade compensation for enhanced flexibility.[28]

While exploring the legal implications of various types of alternative work arrangements is beyond the scope of this report, it is worth noting that one current topic in the field of nonstandard work arrangements concerns the determination of whether workers are considered "employees" with the benefits and legal responsibilities that entails or independent contractors. David Weil's work on the "fissured workplace" explores these issues, and it has been the subject of a number of recent and ongoing legal cases, involving both platform technologies, such as Uber, and more traditional organizations, such as trucking companies.[29]

Types of Nontraditional Work Arrangements

Other models involve allowing employees to shift between part-time and full-time work and to phase into retirement by reducing hours (and compensation) on a gradual basis.[30] According to the National Study of Employers, 41 percent of employers allow at least some employees to move from full-time to part-time work and back again (8 percent allow all or most employees to do so), and 59 percent allow at least some employees to phase into retirement (21 percent allow all or most employees to do so).[31]

---

[27] Tracy Haugen, "Workplaces of the Future: Creating an Elastic Workplace," *Resetting Horizons—Human Capital Trends*, Deloitte, 2013; EY, *Global Generations: A Global Study on Work-Life Challenges Across Generations: Detailed Findings*, 2015; Indeed Hiring Lab, *Targeting Today's Job Seeker: Data, Trends and Insight*, 2017.

[28] Alexandre Mas and Amanda Pallais, "Valuing Alternative Work Arrangements," March 2017.

[29] David Weil, "Lots of Employees Get Misclassified as Contractors: Here's Why IT Matters," *Harvard Business Review*, July 5, 2017.

[30] Leick and Matos, 2017.

[31] Matos, Galinsky, and Bond, 2017.

Terms of Service

An additional category of alternative work arrangement involves flexibility in the terms of service—that is, the legal relationship between the company and the individual performing the work. In the rubrics outlined above that categorize types of nonstandard work arrangement, terms-of-service flexibility would fall under the umbrella of arrangements that involve "limited administrative attachment" (Pfeffer and Baron) or contract work over which the organization for whom the work is done "lacks directive control" (Cappelli and Keller). For example, companies may use independent contractors, day laborers, or temporary services workers in lieu of employing workers directly. According to a Department of Commerce report, the motivation for employers to utilize contractors or temporary workers include to "use labor for shorter periods," "without the cost of offering benefits," and to "free businesses from the high costs of hiring and firing workers."[32] The determination of whether workers are classified as employees or contractors for the purposes of labor law is governed by common law standards.[33] The Department of Labor provided guidance (an "economic realities" test) to interpret these standards in 2015, but this guidance has since been revoked.[34]

Because temporary services are tabulated as a separate industry in BLS data, it is not straightforward to identify which industries use temporary services workers most intensively; however, separate Census Bureau surveys that estimate business spending on temporary workers as a share of total payroll are informative. According to 2012 data, the following industries spend at least 10 percent of payroll costs on tem-

---

[32] Jessica R. Nicholson, "Temporary Help Workers in the U.S. Labor Market," ESA Issue Brief 03-15, Economics and Statistics Administration (ESA), U.S. Department of Commerce, July 1, 2015.

[33] Internal Revenue Service, "Independent Contractor (Self-Employed) or Employee?" undated.

[34] Hodgson Russ LLP, "New U.S. DOL Memo Concludes Most Workers Are Employees, Not Independent Contractors," July 15, 2015; U.S. Department of Labor, "US Secretary of Labor Withdraws Joint Employment, Independent Contractor Informal Guidance," press release, June 7, 2017.

porary workers: telecommunications, pipeline transportation, utilities, warehousing and storage, petroleum and coal products manufacturing, and oil and gas extraction.[35] Data on the occupations of temporary help services industry workers are timelier. Occupational Employment Statistics data from May 2017 indicate that about one-quarter of temporary services workers are in both transportation and material moving occupations and production occupations, with another nearly 20 percent in office and administrative support occupations.[36] At a detailed occupation level, most occupations that comprise at least 1 percent of total temporary services employment are in those three major occupational categories; the exceptions are substitute teachers, HR specialists, construction laborers, registered nurses, janitors and cleaners, and nursing assistants.

Contracting out work (as opposed to utilizing temporary workers) constitutes a related category. According to one recent report,

> [i]ndependent contracting is especially prevalent in industries where:
>
> - Workers move frequently from project to project, or work multiple projects at once;
> - Firms need to be able to respond to short-run changes in demand, or make up for gaps in supply, by calling on more workers than they could economically maintain as traditional employees;
> - It is efficient to be able to evaluate performance, and hence base compensation, on output, as opposed to direct observations of time spent working;
> - There are efficiency benefits to having workers own their own capital (e.g., a truck or taxi).[37]

---

[35] Nicholson, 2015.

[36] BLS, "NAICS 561320—Temporary Help Services," *May 2017 National Industry-Specific Occupational Employment and Wage Estimates*, last modified March 30, 2018c.

[37] Jeffrey A. Eisenach, "The Role of Independent Contractors in the U.S. Economy," Navigant Economics, December 2010, p. i.

According to the CWS fielded in 2017, about 25 percent of independent contractors work in professional and business services, about 20 percent work in construction, and about 10 percent work in financial activities.[38] More than 40 percent of independent contractors listed their occupation as management, professional, or related; nearly 20 percent were in service occupations, and 16 percent were in sales and office occupations. Unlike temporary help services workers, independent contractors are unlikely to be in production, transportation, and material moving occupations.

From a worker's perspective, the rationale for participating in independent work outside of a traditional employer-employee relationship can vary. McKinsey Global Institute has developed a categorization scheme that includes four groups along two axes: whether it is the individual's primary job and the degree of choice over performing this type of alternative work versus traditional work. A "free agent" chooses to be an independent worker and derives the majority of their income from it; a "casual earner" also works in this manner by choice but does it for supplemental income; a "reluctant" would prefer a traditional job yet derives most of their income from independent work; and "the financially strapped" do independent work for supplemental income but would prefer not to have to do the work.[39]

Platform Technologies

Technology plays a major role in expanding the set of tasks that can be performed through flexible work arrangements, as described above, for example, in the case of telework and telemedicine. The Organisation for Economic Co-operation and Development has found that access to flexible work arrangements is "strongly associated with intensive use of information and communication technologies (ICTs) at work, as it facilitates flexible working-time schedules as well as working from

---

[38] BLS, 2018c.

[39] McKinsey Global Institute, *Independent Work: Choice, Necessity, and the Gig Economy*, October 2016.

home."[40] New technologies are also opening the door to entirely new ways of connecting people with work opportunities. Platform technologies that "facilitate direct transactions between consumer and producer" and that provide "flexible schedules for gig workers" are at the core of the gig, or "on-demand," economy.[41] Platforms range from car services (e.g., Uber and Lyft) to food delivery services (e.g., Grubhub and DoorDash) to applications to rent out spare rooms (e.g., Airbnb) to online markets for manual and low-skilled labor (e.g., TaskRabbit) and high-skilled freelance workers (e.g., Upwork). Other new platform technology businesses seek to help workers to patch together a series of gigs to build a "full-time" job (e.g., JobGrouper) or to help companies connect with workers to fill intermittent needs (e.g., Shiftgig and Wonolo).

Some major corporations, including Samsung and other Fortune 500 firms, have expanded their use of platform technologies in recent years to supplement their full-time workforce with freelance workers, in lieu of or in addition to using more "conventional" temporary staffing firms.[42] Reasons for doing, according to one recent report, include to draw in needed expertise "quickly and flexibly," to "lower the startup costs," and to "eliminate or at least reduce geographical, informational, and administrative barriers in the hiring process."[43]

These platforms have been simultaneously heralded as launching a new wave of flexible work that allows people to work on their own time and own terms and criticized as opening the door to exploitation of workers left to fend for themselves without benefits or a safety net. Some policymakers have focused on the "opportunities and challenges" of the "on-demand economy" and have sought to identify how the

---

[40] Organisation for Economic Co-operation and Development, *Be Flexible! Background Brief on How Workplace Flexibility Can Help European Employees to Balance Work and Family*, September 2016.

[41] Emilia Istrate and Jonathan Harris, "The Future of Work: The Rise of the Gig Economy," National Association of Counties, Counties Futures Lab, November 2017.

[42] Greetje F. Corporaal, "Platform Sourcing: How Fortune 500 Firms Are Adopting Online Freelancing Platforms," Oxford Internet Institute, University of Oxford, August 2017.

[43] Corporaal, 2017.

social safety net could evolve in a world in which nonstandard employment becomes increasingly common.[44] A number of economists, business leaders, and others have developed "portable benefits" proposals as a means to bring greater stability to workers.[45] Despite the widespread attention given these platform technologies, estimates in the literature of the share of workers who participate in these work arrangements indicate that "this segment of the workforce is quite small" (less than 1 percent in 2015 to 2016).[46]

Case Studies: Nonstandard Work Arrangements in the Federal Government

Federal agencies have authorities that allow them to reach human capital in ways other than in full-time competitive service positions. We illustrate how two agencies utilize these authorities to surge their skilled civilian workforces during periods of high activity, comparable to the role of the reserve force in DoD. These agencies were of interest because of their ability to offer nonstandard work arrangements to supplement the skills and capacity of a standing workforce (including in one case the practice of working as a team) and the amount of information that was readily available to the project team. The material presented draws from publicly available information supplemented with interviews.[47] Two agencies that have utilized these authorities to provide flexible staffing arrangements to supplement their full-time workforce include SBA's ODA and the NIFC/USFS. Here, we describe

---

[44] Senator Mark R. Warner, "Senator Warner Addresses the Opportunities and Challenges of the 'Sharing Economy,'" webpage, undated.

[45] Aspen Institute, "Portable Benefits," webpage, undated.

[46] Emilie Jackson, Adam Looney, and Shanthi Ramnath, "The Rise of Alternative Work Arrangements: Evidence and Implications for Tax Filing and Benefit Coverage," Office of Tax Analysis, U.S. Department of the Treasury, Working Paper 114, January 2017.

[47] Much of this material was gathered as part of a previous study for the Federal Emergency Management Agency (FEMA) that looked at intermittent surge workforces.

the mechanisms that these agencies use to recruit, retain, and manage their variable workforces.

U.S. Forest Service and National Interagency Fire Center

Wildland firefighting is coordinated through NIFC. NIFC is a consortium of federal agencies with firefighting responsibilities and capabilities. USFS, National Park Service, Bureau of Land Management, U.S. Fish and Wildlife Service, Bureau of Indian Affairs, U.S. Fire Administration, National Oceanic and Atmospheric Administration, and the National Association of State Foresters participate in NIFC. USFS is closely associated with the resourcing and management of NIFC; accordingly, to discern alternative workforce management practices, we treat USFS and NIFC as essentially the same entity.

Firefighting resources are managed according to the militia model; that is, full-time employees of the participating agencies account for 90 percent of staffing, and intermittent staff (both seasonal and casual employees) make up the remaining 10 percent. In the event of a fire, local resources are employed initially. As the severity and complexity of the fire increases, resources are brought in from an expanding geographic area. There are five levels of incident response, the smallest of which would draw from firefighters in a small surrounding community (type 5) to the most complex (type 1).[48]

USFS hires two classes of intermittent surge personnel: casual and seasonal.

---

[48] For reference, type 3 teams coordinate a large metropolitan area or statewide response to an incident. Type 2 and type 1 incidents are the largest and most complex; they draw standing national management teams to provide the command and control infrastructure needed to plan and manage the operational, logistical, safety, and community issues related to the incident. Nationwide, there are 35 type 2 and 17 type 1 incident management teams. See National Interagency Fire Center, "National Interagency Incident Management Team Rotation," November 2018.

Casual Personnel

Casual personnel are hired under an administratively determined pay plan mechanism and are paid only when activated.[49] They may be hired when there is a fire, natural disaster, or other emergency with the potential to cause loss of life, serious injury, or damage to federally protected property or natural or cultural resources; to provide assistance to states participating in various formal agreements; or to meet mission responsibilities as assigned by FEMA. Hiring is for an uncertain and temporary period and must be terminated when other employment methods can be used. Rates of pay are determined annually by the National Wildfire Coordinating Group Incident Business Committee with input from officials representing the geographic areas.[50] Those hired are generally retired civil servants who have firefighting qualifications. Once their qualifications are verified, they can be activated on a situational basis. Casual personnel are hired locally and managed by the geographic areas. While there are some inconsistencies among the areas, our interviewee felt it was more important that casuals are managed by the people who are personally aware of their training levels and overall capabilities. Casuals do not receive other benefits or training.

Seasonal Employees

Seasonal employees (referred to as 1039s) are frequently younger people looking to gain experience and develop their skills. The USFS will often provide some career development training for this class of surge worker. Seasonal employees are hired noncompetitively using the authority provided in 5 U.S.C. 2103 (see 5 CFR 213.104 for the conditions). The guidance states that these employees may not work more

---

[49] Administratively determined plans are pursuant to 5 U.S.C. 5102(c)(19), 7 U.S.C. 2225 and 2226, 16 U.S.C. 554e, and 43 U.S.C. 1469. Persons hired as casual firefighters must be at least 18 years old; meet minimum physical fitness standards as established by agency policy; meet minimum training requirements for the position before assignment; fulfill agency security requirements; have proper clothing and footgear; and, if in a unit leadership position, must be proficient in the English language and the language used by members of their units. See National Wildfire Coordinating Group, *NWCG Standards for Interagency Incident Business Management*, Boise, Id., PMS 902, April 2018.

[50] National Wildfire Coordinating Group, 2018.

than 1,040 hours in a year (excluding overtime and 80 hours of eligible training).

The ODA works in parallel with FEMA during declared disasters to provide low-cost loans to eligible households and businesses. ODA employees must be "available to respond to disaster emergencies anywhere in the United States and its territories, often within 48 hours or less." The ODA has some flexibility to manage disaster workloads among their call center locations. However, should disaster activity exceed the capability of their core staff, the ODA hires surge staff for temporary, intermittent work. Surge personnel are hired in several different classifications; the two most commonly used are ODA's

- *Surge personnel.* Term-intermittent personnel hired on a competitive basis,[51] surge personnel are term-limited employees who are placed on a roster and who expect to work on an on-call basis in either the field or operations center. (About 90 percent of ODA employees work in an operations center.)
- *Surge plus personnel.* Excepted personnel hired as needed, these at-will employees are hired only when other personnel categories are insufficient to meet demand. These are noncompetitive positions under SBA's exemption for a period of two or four years depending on prior service.[52]

---

[51] See 5 CFR 340.403. Term intermittent employees are hired to work intermittently on an on-call basis. Hired competitively, their terms can be between one and four years. These employees work only if there is increased disaster activity or specific business needs; must report to work when called in accordance with the employment agreement; are trained as needed; and generally are not eligible for full benefits. SBA, "Office of Disaster Assistance Staffing Strategy," June 4, 2014.

[52] See 5 CFR 213.104; and OPM, 2014. According to the ODA Staffing Strategy, surge plus personnel are called when the core and surge staff capability has been surpassed. When needed, individuals are hired for a period of four years or two years using SBA's special excepted service hiring authority. This authority may be used when there is a critical need for a short-term job, for temporary work in a remote or isolated location, to hire a noncitizen because no qualified citizen is available, or when specialized skills are needed. SBA, 2014.

As of May 2014, surge and surge plus staff totaled just under 3,000.[53] At activation, employees are given an expectation for the likely deployment length, which is generally 40 to 60 days.

A major feature of the ODA staffing strategy is to leverage the Disaster Field Operations Centers—call centers based in Sacramento (for disasters west of the Mississippi) and Atlanta (for disasters east of the Mississippi). Since about 90 percent of the loans made by ODA are processed through the call centers using electronic applications, call centers can support each other in managing workloads. These call centers report directly to the associate administrator for ODA.

Organizational Structure

Within ODA, the headquarters is involved in recruiting, hiring, and deploying surge personnel, but its two Disaster Field Operations Centers initiate and manage activation requests according to conditions established by ODA and subject to ODA approval. Each operations center maintains a surge personnel roster, and ODA maintains a master list.[54] In contrast, the military RCs are managed at the unit or local team level with regional and national oversight. The team leaders of distributed, deployable teams are responsible for recruiting and hiring, as well as for meeting training and operational standards.

Comparison of Recruiting and Hiring Practices Across the Military Reserve Components, Office of Disaster Assistance, and U.S. Forest Service

Most agencies hire on a much smaller scale than DoD and require individuals to have specific qualifications, levels of experience, or certifications. Unlike the military RCs, surge personnel are hired only if they are qualified to perform the job with relatively little additional training. The most common recruiting methods are the USAJOBS website (for U.S. government agencies) and word of mouth, which provides

---

[53] SBA, 2014.

[54] SBA, 2014.

a way to tap into communities of practice to fill surge needs.[55] ODA recruits loan officers, lawyers, and other practitioners from the mortgage lending industry. USFS/NIFC requires firefighters to have an incident qualifications card. The military RCs recruit and train some personnel without prior military experience; however, many of their recruits were previously on active duty and already have the necessary skills. By tapping into communities of practice, these programs do not have to invest as much in training and can readily expand their workforces when necessary. Communities of practice can also be a source of organizational innovation when staff from those communities are familiar with emerging or best practices in their respective fields.

Table C.2 presents the hiring mechanisms, reasons for hiring surge personnel, qualifications required, and communities of practice.

Comparison of Training and Force Development Models

ODA and USFS only hire people as surge employees who already possess specific certifications, technical qualifications, or levels of experience into well-defined positions that are comparable to private-sector jobs and national certifications (red card). For example, there are nine different job specialties listed on the SBA website in the areas of legal, construction, loan application assistance and processing, and IT.[56] Because it hires from industries with comparable skills and has automated much of the process, ODA minimizes training requirements, although some basic level is provided either centrally or on the job. ODA also relies heavily on cross-training individuals to perform several different functions, so the workforce is used more efficiently. Training is provided at a central location, and individuals who are hired on short notice during a disaster can be trained on the job.

---

[55] Communities of practice are defined as "groups of people who share a concern, a set of problems, or a passion about a topic, and who deepen their knowledge and expertise in this area by interacting on an ongoing basis." John L. Cordery, Edward Cripps, Cristina B. Gibson, Christine Soo, Bradley L. Kirkman, and John E. Mathieu, "The Operational Impact of Organizational Communities of Practice: A Bayesian Approach to Analyzing Organizational Change," *Journal of Management*, Vol. 41, No. 2, 2014, p. 7.

[56] SBA, 2014.

Table C.2
Hiring Mechanisms and Qualification Requirements of Examined Programs

Program	Hiring Mechanisms	Purposes of Surge	Requirements	Community of Practice
Military RCs	Recruit and train; flow from AC	Major wartime surge; participation in steady-state deployments	Age, vocational aptitude, and physical standards	Current and former AC members
USFS/NIFC	Casuals	Rapid response to fires	Incident qualification card (red card)	State and local emergency response
	Seasonal employees	Surge response to fires	Training, experience, and fitness standards	State and local emergency response
ODA	Term-intermittent	Disaster response	Paralegals, loan officers, etc.	Commercial mortgage industry
	At will and contractors	Disaster response	Paralegals, loan officers, etc.	Commercial mortgage industry

Both ODA and USFS use at-will employees for meeting peak demands and do not generally provide additional training, while seasonal employees may receive some limited training. For example, ODA will attempt to deploy seasonal hires in positions that offer professional growth and development. USFS/NIFC will invest in some training for the seasonal employees but not for casuals. It has been noted that it can be challenging for intermittent employees to stay up to date on policy and procedural changes.

Comparison of Activation Practices

ODA rotates through the roster activating staff according to length since the last deployment, cumulative days deployed, and (if applicable) professional development opportunities. Seasonals and surge staff are activated individually, although surge personnel work for a handful of operations centers, where they likely develop working relationships. ODA has a three-declinations policy that is typically enforced.

USFS calls up those in the affected area first. Wildland firefighters are typically deployed from 14 days up to 30 days, exclusive of travel from and to their home units. Opportunities for deployment have been plentiful in recent years. Firefighting resources are applied according to the scale and severity of the event. The geographic region takes the lead for smaller-scale fires. When a geographic area's resources are fully allocated, NIFC reaches out to other regional and national teams. When there is competition for firefighting resources nationally, the National Interagency Coordinating Center distributes resources using an established prioritization (protect human life, infrastructure, etc.). In addition to local crews, national teams are in place to provide highly skilled and experienced management capability for large, complex events. Two levels of national teams, which include a mix of full-time and surge personnel (seasonals and casuals), are placed on alert on a rotating schedule. When there is an event, the host agency performs a complexity analysis that will generate a suggested management level. If it is a type 1 incident (the most severe), 27 to 47 people will deploy in a type 1 national team. Depending on negotiations with the host agency, the deployed team includes all personnel types (full time, seasonal, casual), as well as both qualified and trainee personnel. Type 2 teams deploy using the same process but with smaller contingents. Depending on their specific qualifications, casuals have plentiful opportunities to deploy. In periods of lesser fire activity, NIFC tries to avoid using seasonals or casuals; as a result, seasonals may not get needed experience-based qualifications as quickly, slowing their career progression. USFS noted that declinations were generally handled informally, especially when the surge personnel are known to program managers and circumstantial factors are considered.

Comparison of Compensation Levels
ODA and USFS provide compensation for hours worked during deployment. USFS pays travel, permanent personnel hazard pay, and seasonals get overtime pay, but casuals receive neither hazard nor overtime pay. Pay rates for casuals are administratively determined, and, in 2016, hourly rates for the 13 distinct classification levels ranged from

just under $15 per hour to over $56 per hour.[57] Most firefighters are housed in camps and therefore do not get per diem. In contrast, the military RCs pay a lot for training and for the time reservists spend in training.

Workforce Requirements

ODA has a disaster-resourcing spreadsheet model that develops estimates of monthly staffing needs using operational plans and estimates of loan applications based on FEMA activity. ODA's model for determining the resources needed for an incident uses multiple estimates as inputs:

- number of offices and working hours
- number of loan applications
- split between homes and businesses
- duration of the disaster and a time-phased distribution of loan applications (based on the event size and historical experience)
- productivity factors (number of files processed/hour).

The USFS/NIFC has a typing model for determining the level of management team required for an event. The host agency will perform a complexity analysis to determine a suggested incident management level, and firefighting crews are sized according to such factors as acreage burning. When there is a wildland fire, the affected geographic area headquarters coordinates the resources to fight the fire and contacts the National Interagency Coordination Center if additional resources are needed. If national resources are constrained, the National Interagency Coordination Center can also reach out to international partners. The host agency can negotiate for up to 20 additional people on these teams. A National Interagency Mobilization Guide identifies standard procedures for these teams.

---

[57] USFS, *Interagency Incident Business Management Handbook*, FSH 5109.34, April 1, 2017.

References

"70% des réservistes sans passé militaire" ["70% of Reservists Do Not Have a Military History"], *Le Courrier de l'ouest*, April 30, 2018.

AEI-Brookings Working Group on Paid Family Leave, *Paid Family and Medical Leave: An Issue Whose Time Has Come*, Washington, D.C., May 2017. As of July 11, 2018:
https://www.brookings.edu/wp-content/uploads/2017/06/es_20170606_paidfamilyleave.pdf

Air Force Instruction 36-2906, *Personal Financial Responsibility*, July 30, 2018.

ArmedForces.co.uk, "Reserve Forces: Types of Reservist and Manpower Strength," webpage, undated. As of May 3, 2018:
http://www.armedforces.co.uk/army/listings/l0070.html

Armée de Terre, *Création du bataillon de reserve Ile de France—24e régiment d'infanterie à Vincennes*, July 2, 2013. As of March 29, 2019: https://www.youtube.com/watch?v=c4mF1YYki84

Army Regulation 670-1, *Uniform and Insignia Wear and Appearance of Army Uniforms and Insignia*, Washington, D.C., March 31, 2014.

Army Regulation 165-1, *Religious Activities: Army Chaplain Corps Activities*, Washington, D.C., June 12, 2015.

Army Regulation 601-210, *Regular Army and Reserve Components Enlistment Program*, Washington, D.C., August 31, 2016.

Army Regulation 135-18, *The Active Guard Reserve Program*, Washington, D.C., September 29, 2017.

Army Study Guide, "Hair Standards," webpage, undated. As of October 2, 2018:
https://www.armystudyguide.com/content/Prep_For_Basic_Training/Prep_for_basic_uniforms/hair-standards.shtml

Asch, Beth J., Michael G. Mattock, and James Hosek, *The Blended Retirement System: Retention Effects and Continuation Pay Cost Estimates for the Armed Services*, Santa Monica, Calif.: RAND Corporation, RR-1887-OSD/USCG, 2017. As of March 13, 2019: https://www.rand.org/pubs/research_reports/RR1887.html

Ashwood, J. Scott, Ateev Mehrotra, David Cowling, and Lori Uscher-Pines, "Direct-to-Consumer Telehealth May Increase Access to Care but Does Not Decrease Spending," *Health Affairs*, Vol. 36, No. 3, 2017, pp. 485–491.

Aspen Institute, "Portable Benefits," webpage, undated. As of July 11, 2018: https://www.aspeninstitute.org/programs/future-of-work/portable-benefits

Association des Officers de Réserve des Corps de l'Armement, "La réserve militaire DGA," *AORCA: Association des Officiers de Réserve des Corps de l'Armement* (blog), undated. As of March 29, 2019: https://aorca.fr/public/index.php/la-reserve-dga/

Australian Department of Defence, "ADF Total Workforce Model," webpage, undated(a). As of November 8, 2018: http://www.defence.gov.au/ADF-TotalWorkforceModel

———, "The Service Spectrum," webpage, undated(b). As of November 8, 2018: http://www.defence.gov.au/ADF-TotalWorkforceModel/ServiceSpectrum.asp

———, "Work Life Balance," webpage, undated(c). As of October 2, 2018: http://www.defence.gov.au/APSCareers/What-we-offer/work-life-balance.asp

———, *ADF Total Workforce Model Frequently Asked Questions*, version 3.0, Canberra, 2018. As of May 12, 2018: http://www.defence.gov.au/ADF-TotalWorkforceModel/Docs/181130-ADFTWM -Frequently-Asked-Questions.pdf

Australian National War Memorial, "Conscription," *Memorial Encyclopedia*, last updated October 23, 2018. As of November 12, 2018: https://www.awm.gov.au/articles/encyclopedia/conscription

Barger, Laura K., Shantha M. W. Rajaratnam, Wei Wang, Conor S. O'Brien, Jason P. Sullivan, Salim Qadri, Steven W. Lockley, and Charles Czeisler, "Common Sleep Disorders Increase Risk of Motor Vehicle Crashes and Adverse Health Outcomes in Firefighters," *Journal of Clinical Sleep Medicine*, Vol. 11, No. 3, March 2015, pp. 233–240.

Beaghen, Michael, and Lynn Weidman, *Statistical Issues of Interpretation of the American Community Survey's One-, Three-, and Five-Year Period Estimates*, Washington, D.C.: U.S. Census Bureau, 2008. As of November 9, 2018: https://usa.ipums.org/usa/resources/MYE_Guidelines.pdf

BLS—*See* Bureau of Labor Statistics.

Brown, Gary, *Military Conscription: Issues for Australia*, Australian Parliament, Foreign Affairs, Defence and Trade Group, October 12, 1999. As of November 12, 2018:
https://www.aph.gov.au/About_Parliament/Parliamentary_Departments /Parliamentary_Library/Publications_Archive/CIB/cib9900/2000CIB07

Brunelli, Laureen Miles, "Medical Jobs from Home," *The Balance Careers*, June 9, 2017. As of July 11, 2018:
https://www.thebalancecareers.com/medical-jobs-from-home-3542685

Bureau of Labor Statistics, U.S. Department of Labor, "Census 2010 Occupation Codes," undated(a). As of October 2, 2018:
https://www.bls.gov/tus/census10ocodes.pdf

———, "Employment Projections," undated(b). As of October 30, 2018:
https://data.bls.gov/projections/occupationProj

———, "Industries with Summer Employment Peaks," *Economics Daily*, July 15, 2014a. As of October 28, 2018:
https://www.bls.gov/opub/ted/2014/ted_20140715.htm

———, "Industries with Winter Employment Peaks," *Economics Daily*, December 3, 2014b. As of October 28, 2018:
https://www.bls.gov/opub/ted/2014/ted_20141203.htm

———, "Employed Persons by Detailed Occupation, Sex, Race, and Hispanic or Latino Ethnicity," *Labor Force Statistics from the Current Population Survey*, data as of January 19, 2018a. As of July 10, 2018:
https://www.bls.gov/cps/

———, "Persons at Work by Occupation, Sex, and Usual Full- or Part-Time Status," *Labor Force Statistics from the Current Population Survey,* last modified January 19, 2018b. As of July 10, 2018:
https://www.bls.gov/cps/cpsaat23.htm

———, "NAICS 561320—Temporary Help Services," *May 2017 National Industry-Specific Occupational Employment and Wage Estimates*, last modified March 30, 2018c. As of July 11, 2018:
https://www.bls.gov/oes/current/naics5_561320.htm

———, "Contingent and Alternative Employment Arrangements—May 2017," press release, USDL-18-0942, June 7, 2018d. As of July 3, 2018:
https://www.bls.gov/news.release/pdf/conemp.pdf

———, "American Time Use Survey—2017 Results," press release, USDL-18-1058, June 28, 2018e. As of July 10, 2018:
https://www.bls.gov/news.release/pdf/atus.pdf

———, "Employed Persons Working on Main Job and Time Spent Working on Days Worked by Class of Worker, Occupation, Earnings, and Day of Week, 2017 Annual Averages," press release, June 28, 2018f. As of October 2, 2018:
https://www.bls.gov/news.release/atus.t05.htm

Burke, Robert P., Chief of Naval Personnel, U.S. Navy, testimony before the U.S. House Armed Services Committee, Subcommittee on Military Personnel, May 17, 2017. As of October 2, 2018:
https://plus.cq.com/doc/testimony-5103052?0

———, testimony before the U.S. House Armed Services Committee, Subcommittee on Military Personnel, April 13, 2018. As of October 2, 2018:
https://docs.house.gov/meetings/AS/AS02/20180413/108074/HHRG-115-AS02
-Wstate-BurkeR-20180413.pdf

California Military Department, "California State Military Reserve," webpage, undated. As of November 12, 2018: https://calguard.ca.gov/csmr

Caminiti, Susan, "4 Gig Economy Trends That Are Radically Transforming the US Job Market," CNBC.com, October 29, 2018. As of February 18, 2019:
https://www.cnbc.com/2018/10/29/4-gig-economy-trends-that-are-radically
-transforming-the-us-job-market.html

Cappelli, Peter, and J. R. Keller, "Classifying Work in the New Economy," *Academy of Management Review*, Vol. 38, No. 4, 2013, pp. 575–596.

———, "Talent Management: Conceptual Approaches and Practical Challenges," *Annual Review of Organizational Psychology and Organizational Behavior*, No. 1, March 2014, pp. 305–331.

Carafano, James Jay, and John R. Brinkerhoff, *Katrina's Forgotten Responders: State Defense Forces Play a Vital Role*, Washington, D.C.: Heritage Foundation, October 5, 2005. As of November 8, 2018:
https://www.heritage.org/homeland-security/report/katrinas-forgotten-responders
-state-defense-forces-play-vital-role

Carter, Phillip, Katherine Kidder, Amy Schafer, and Andrew Swick, *AVF 4.0: The Future of the All-Volunteer Force*, working paper, Washington, D.C.: Center for a New American Security, March 28, 2017. As of November 8, 2018:
https://www.cnas.org/publications/reports/avf-4-0-the-future-of-the-all-volunteer
-force

Cawley, John, and Catherine Maclean, "Unfit for Service: The Implications of Rising Obesity for US Military Recruitment," *Health Economics*, Vol. 21, No. 11, November 2012, pp. 1348–1366.

Center for Applied Research in the Apostolate, "Frequently Requested Church Statistics," webpage, undated. As of October 2, 2018:
https://cara.georgetown.edu/frequently-requested-church-statistics

Center for State and Local Government Excellence, *Survey Findings: State and Local Government Workforce; 2018 Data and 10 Year Trends*, Washington, D.C., May 2018. As of July 19, 2018:
https://slge.org/resources/state-and-local-government-workforce-2018-data-and-10
-year-trends

Central Intelligence Agency, "Australia," *The World Factbook*, last updated November 6, 2018. As of February 19, 2019:
https://www.cia.gov/library/Publications/the-world-factbook/geos/as.html

———, "Estonia," *The World Factbook*, last updated October 23, 2018. As of February 19, 2019:
https://www.cia.gov/Library/publications/the-world-factbook/geos/en.html

———, "France," *The World Factbook*, last updated October 23, 2018. As of February 19, 2019:
https://www.cia.gov/library/publications/the-world-factbook/geos/fr.html

———, "United Kingdom," *The World Factbook*, last updated October 23, 2018. As of February 19, 2019:
https://www.cia.gov/library/publications/the-world-factbook/geos/uk.html

Chapman, Dennis P., *Manning Reserve Component Units for Mobilization: Army and Air Force Practice*, Land Warfare Paper No. 74, Washington, D.C.: Association of the United States Army, September 2009. As of November 8, 2018:
https://www.ausa.org/sites/default/files/LWP-74-Manning-Reserve-Component-Units-for-Mobilization-Army-and-Air-Force-Practice.pdf

Chong, Yinong, Cheryl D. Fryar, and Qiuping Gu, "Prescription Sleep Aid Use Among Adults: United States, 2005–2010," National Center for Health Statistics Data Brief No. 127, August 2013. As of October 2, 2018:
https://www.cdc.gov/nchs/data/databriefs/db127.htm

Clever, Molly, and David R. Segal, "The Demographics of Military Children and Families," *The Future of Children,* Vol. 23, No. 2, Fall 2013, pp. 13–39.

Colquitt, Jason, Jeffrey Lepine, and Michael Wesson, *Organizational Behavior: Improving Performance and Commitment in the Workplace*, 4th ed., New York: McGraw-Hill Education, 2015.

Comenetz, Joshua, "Census-Based Estimation of the Hasidic Jewish Population," *Contemporary Jewry*, Vol. 26, 2006, pp. 35–74.

Cordery, John L., Edward Cripps, Cristina B. Gibson, Christine Soo, Bradley L. Kirkman, and John E. Mathieu, "The Operational Impact of Organizational Communities of Practice: A Bayesian Approach to Analyzing Organizational Change," *Journal of Management*, Vol. 41, No. 2, 2014, pp. 644–664.

Corporaal, Greetje F., "Platform Sourcing: How Fortune 500 Firms Are Adopting Online Freelancing Platforms," Oxford Internet Institute, University of Oxford, August 2017. As of July 5, 2018:
https://www.oii.ox.ac.uk/publications/platform-sourcing.pdf

Cotton, Sarah K., Ulrich Petersohn, Molly Dunigan, Q. Burkhart, Megan Zander Cotugno, Edward O'Connell, and Michael Webber, *Hired Guns: Views About Armed Contractors in Operation Iraqi Freedom*, Santa Monica, Calif.: RAND Corporation, MG-987-SRF, 2010. As of August 24, 2018:
https://www.rand.org/pubs/monographs/MG987.html

Council of Economic Advisers, *The Economics of Paid and Unpaid Leave*, Washington, D.C.: Executive Office of the President, June 2014a. As of July 11, 2018:
https://obamawhitehouse.archives.gov/sites/default/files/docs/leave_report_final.pdf

———, *Work-Life Balance and the Economics of Workplace Flexibility*, Washington, D.C.: Executive Office of the President, June 2014b. As of July 3, 2018:
https://obamawhitehouse.archives.gov/sites/default/files/docs/updated_workplace_flex_report_final_0.pdf

Cuhls, Kerstin, ed., *Delphi Surveys: Teaching Material for UNIDO Foresight Seminars*, Vienna, Austria: United Nations Industrial Development Organization, 2005.

Czosseck, Christian, Rain Ottis, and Anna-Maria Talihärm, *Estonia After the 2007 Cyber Attacks: Legal, Strategic, and Organisational Challenges in Cyber Security*, Tallinn, Estonia: Cooperative Cyber Defense Centre of Excellence, undated. As of November 12, 2018:
https://pdfs.semanticscholar.org/174b/4bd588306e3fa0067eda83df61719a0996ce.pdf

Davies, Andrew, and Hugh Smith, *Stepping Up: Part-Time Forces and ADF Capability*, Australia Strategic Policy Institute, Strategic Insights No. 44, November 2008. As of November 8, 2018:
https://www.aspi.org.au/report/strategic-insights-44-stepping-part-time-forces-and-adf-capability

Defense Manpower Data Center, "Number of Military and DoD Appropriated Fund (APF) Civilian Personnel Permanently Assigned," webpage, June 30, 2018. As of October 2, 2018:
https://www.dmdc.osd.mil/appj/dwp/dwp_reports.jsp

Dennett, Harley, "Giving a Commitment Choice with the 'Service Spectrum,'" *The Mandarin*, April 20, 2018. As of October 2, 2018:
https://www.themandarin.comau/63500-total-workforce-giving-employees-commitment-choice-service-spectrum

Deputy Chief of Staff of the Army, G-1, *Army Mobilization and Deployment Reference (AMDR)*, last updated January 18, 2018. As of August 9, 2018:
http://www.armyg1.army.mil/documents/AMDR180116.pdf

Desilver, Drew, "Access to Paid Family Leave Varies Widely Across Employers, Industries," *Fact Tank* (blog), Pew Research Center, March 23, 2017. As of July 11, 2018:
http://www.pewresearch.org/fact-tank/2017/03/23/access-to-paid-family-leave
-varies-widely-across-employers-industries

DeZube, Dona, "Seasonal Tax-Preparation Jobs," *Career Advice* (blog), Monster .com, undated. As of July 10, 2018:
https://www.monster.com/career-advice/article/seasonal-tax-preparation-jobs

Direction de l'information légale et administrative (Premier ministre) [Directorate of Legal and Administrative Information, Office of the Prime Minister], "Réserve opérationnelle" ["Operational Reserve"], August 23, 2018. As of November 8, 2018:
https://www.service-public.fr/particuliers/vosdroits/F1188

Dixon, Lauren, "How Ryan Flipped to Flex," *Talent Economy*, December 14, 2016. As of July 10, 2018:
http://www.talenteconomy.io/2016/12/14/ryan-flipped-flex

DoD—*See* U.S. Department of Defense.

"DoD Projects Growth in Civilian Workforce," *FEDweek*, June 5, 2018. As of October 3, 2018:
https://www.fedweek.com/fedweek/dod-projects-growth-in-civilian-workforce/

DoDI—*See* U.S. Department of Defense Instruction.

Dorsey, E. Ray, and Eric J. Topol, "State of Telehealth," *New England Journal of Medicine*, Vol. 375, No. 2, July 14, 2016, pp. 154–161.

Dunigan, Molly, "The Future of US Military Contracting: Current Trends and Future Implications," *International Journal*, Vol. 69, No. 4, 2014, pp. 510–524.

Dunigan, Molly, Carrie M. Farmer, Rachel M. Burns, Alison Hawks, and Claude Messan Setodji, *Out of the Shadows: The Health and Well-Being of Private Contractors Working in Conflict Environments*, Santa Monica, Calif.: RAND Corporation, RR-420-RC, 2013. As of July 19, 2018:
https://www.rand.org/pubs/research_reports/RR420.html

Dunigan, Molly, Susan S. Sohler Everingham, Todd Nichols, Michael Schwille, and Susanne Sondergaard, *Expeditionary Civilians: Creating a Viable Practice of Department of Defense Civilian Deployment*, Santa Monica, Calif.: RAND Corporation, RR-975-OSD, 2016. As of July 19, 2018:
https://www.rand.org/pubs/research_reports/RR975.html

Edmunds, Timothy, Antonia Dawes, Paul Higate, K. Neil Jenkings, and Rachel Woodward, "Reserve Forces and the Transformation of British Military Organisation: Soldiers, Citizens and Society," *Defence Studies*, Vol. 16, No. 2, 2016, pp. 118–136.

Eggleton, Mark, "Australian Armed Forces Face Demographic Crisis," *Financial Review*, March 16, 2015. As of November 12, 2018:
https://www.afr.com/news/special-reports/defence-and-national-security/australian-armed-forces-face-demographic-crisis-20150311-141faq

Eisenach, Jeffrey A., "The Role of Independent Contractors in the U.S. Economy," Navigant Economics, December 2010. As of July 11, 2018:
https://iccoalition.org/wp-content/uploads/2014/07/Role-of-Independent-Contractors-December-2010-Final.pdf

Ell, Kellie, "Want a Holiday Job? Here's Who's Hiring Seasonal Employees," *USA Today*, October 25, 2017. As of July 10, 2018:
https://www.usatoday.com/story/money/2017/10/11/want-holiday-job-heres-whos-hiring-seasonal-employees/724912001

England, Gordon, Deputy Secretary of Defense, "Implementation of Section 324 of the National Defense Authorization Act for Fiscal Year 2008 (FY 2008 NDAA)—Guidelines and Procedures on In-Sourcing New and Contracted Out Functions," memorandum, April 4, 2008. As of July 19, 2018:
http://prhome.defense.gov/Portals/52/Documents/RFM/TFPRQ/Docs/OSD%20In-sourcing%20Guidance%2004184-08.pdf

Estonian Defence Forces, "Compulsory Military Service," webpage, last updated March 11, 2014a. As of May 5, 2018:
http://www.mil.ee/en/defence-forces/compulsory-military-service

———, "Reserve Force," webpage, last updated March 11, 2014b. As of May 5, 2018:
http://www.mil.ee/en/defence-forces/Reserve-Force

Estonian Defence League Act, February 28, 2013. As of May 5, 2018:
https://www.riigiteataja.ee/en/eli/ee/Riigikogu/act/521032014005/consolide

Estonian Military Service Act, June 12, 2012. As of October 3, 2018:
https://www.riigiteataja.ee/en/eli/513032017006/consolide

EY, *Global Generations: A Global Study on Work-Life Challenges Across Generations, Detailed Findings*, 2015. As of July 10, 2018:
https://www.ey.com/Publication/vwLUAssets/Global_generations_study/$FILE/EY-global-generations-a-global-study-on-work-life-challenges-across-generations.pdf

Fatsis, Stefan, "What Patriotism Means to the NFL," *Slate*, May 31, 2018. As of November 12, 2018:
https://slate.com/culture/2018/05/how-the-nfl-helped-players-dodge-the-draft-during-the-vietnam-war.html

Federal Aviation Administration, "U.S. Civil Airmen Statistics," webpage, February 16, 2018. As of October 2, 2018:
https://www.faa.gov/data_research/aviation_data_statistics/civil_airmen_statistics

Feds Hire Vets, "Special Authorities for Hiring Veterans," webpage, OPM, undated. As of October 2, 2018:
https://www.fedshirevets.gov/job-seekers/special-hiring-authorities

Finn, Dennis, and Anne Donovan, *PwC's NextGen: A Global Generational Study, 2013; Summary and Compendium of Findings*, PwC, 2013. As of July 12, 2018:
https://www.pwc.com/gx/en/hr-management-services/pdf/pwc-nextgen-study-2013.pdf

Flegal, Katherine M., Deanna Kruszon-Moran, Margaret D. Carroll, Cheryl D. Fryar, and Cynthia L. Ogden, "Trends in Obesity Among Adults in the United States, 2005 to 2014," *Journal of the American Medical Association*, Vol. 315, No. 21, 2016, pp. 2284–2291.

Flexjobs.com, "Doctor and Practitioner Remote, Part-Time, and Freelance Jobs," webpage, undated(a). As of July 11, 2018:
https://www.flexjobs.com/jobs/doctor-practitioner

———, "State and Local Government Remote, Part-Time, and Freelance Jobs," webpage, undated(b). As of July 19, 2018:
https://www.flexjobs.com/jobs/state-local-government.html

Ford, Earl S., Anne G. Wheaton, Timothy J. Cunningham, Wayne H. Giles, Daniel P. Chapman, and Janet B. Croft, "Trends in Outpatient Visits for Insomnia, Sleep Apnea, and Prescriptions for Sleep Medications Among US Adults: Findings from the National Ambulatory Medical Care Survey 1999–2010," *Sleep*, Vol. 37, No. 8, 2014, pp. 1283–1293.

GAO—*See* U.S. Government Accountability Office.

Gauthier, Padre, "Diaconal Council," trans. Padre Delisle, Ottawa: Roman Catholic Military Ordinariate of Canada, undated. As of November 12, 2018:
http://www.rcmilord.com/main.php?p=82

Gautier, Mary L., and Thomas P. Gaunt, *A Portrait of the Permanent Diaconate: A Study for the U.S. Conference of Catholic Bishops, 2014–2015*, Washington, D.C.: Center for Research in the Apostolate, Georgetown University, May 2015. As of November 12, 2018:
http://www.usccb.org/beliefs-and-teachings/vocations/diaconate/upload/Diaconate-Post-ordination Report-2015-FINAL.pdf

Gelt, Jessica, "The Latest Trend Among Chefs: Food Tattoos," *Los Angeles Times*, April 27, 2012.

George, Elizabeth, and Prithviraj Chattopadhyay, *Non-Standard Work and Workers: Organizational Implications*, Geneva, Switzerland: International Labour Office, Conditions of Work and Employment Series No. 61, October 14, 2015. As of July 10, 2018:
http://www.ilo.org/travail/whatwedo/publications/WCMS_414581/lang—en/index.htm

————, "Understanding Nonstandard Work Arrangements: Using Research to Inform Practice," Society for Human Resource Management and Society for Industrial and Organizational Psychology, 2017. As of July 3, 2018: http://www.siop.org/Portals/84/docs/SIOP-SHRM%20White%20Papers/2017 _03_SHRM-SIOP_Nonstandard_Workers.pdf

Georgetown Federal Legislation Clinic, "The Federal Employees Flexible and Compressed Work Schedules Act (FEFCWA)," Workplace Flexibility 2010, Georgetown University Law Center, 2006. As of July 17, 2018: https://scholarship.law.georgetown.edu/cgi/viewcontent.cgi?article=1014&context =legal

Golden, Lonnie, "Irregular Work Scheduling and Its Consequences," Washington, D.C.: Economic Policy Institute, Briefing Paper 394, April 2015. As of July 10, 2018: https://www.epi.org/publication/irregular-work-scheduling-and-its-consequences

Government Publication Office, "John S. McCain National Defense Authorization Act for Fiscal Year 2019," January 3, 2018.

Grandner, Michael A., Jennifer L. Martin, Nirav P. Patel, Nicholas J. Jackson, Philip R. Gehrman, Grace Pien, Michael L. Perlis, Dawei Xie, Daohang Sha, Terri Weaver, and Nalaka S. Gooneratne, "Age and Sleep Disturbances Among American Men and Women: Data from the U.S. Behavioral Risk Factor Surveillance System," *Sleep*, Vol. 35, No. 3, March 1, 2012, pp. 395–406.

Greenfield, Stuart, "Public Sector Employment: The Current Situation," Center for State and Local Government Excellence, undated. As of July 19, 2018: https://slge.org/resources/public-sector-employment-the-current-situation

Greenhouse, Steven, "The Age Premium: Retaining Older Workers," *New York Times*, May 14, 2014.

Guibert, Nathalie, "Ruée des jeunes Français vers les armées," *Le Monde*, November 20, 2015. As of May 2, 2019: https://www.lemonde.fr/attaques-a-paris/article/2015/11/19/ruee-des-jeunes -francais-vers-les-armees_4813438_4809495.html

Hall, Kimberly Curry, Margaret C. Harrell, Barbara Bicksler, Robert Stewart, and Michael P. Fisher, *Veteran Employment: Lessons from the 100,000 Jobs Mission*, Santa Monica, Calif.: RAND Corporation, RR-836-JPMCF, 2014. As of August 21, 2018: https://www.rand.org/pubs/research_reports/RR836.html

H&R Block, "A Little About H&R Block," fact sheet, 2017a. As of July 10, 2018: https://www.hrblock.com/corporate/pdfs/hrb-factsheet.pdf

————, *2017 Annual Report*, Kansas City, Mo., 2017b. As of July 10, 2018: http://investors.hrblock.com/static-files/954d81c4-c732-4dab-a37b-77e994c3437a

Harrell, Margaret C., and Nancy Berglass, *Employing America's Veterans: Perspectives from Businesses*, Washington, D.C.: Center for a New American Security, June 2012. As of November 12, 2018:
https://www.cnas.org/publications/reports/employing-americas-veterans
-perspectives-from-businesses

Haugen, Tracy, "Workplaces of the Future: Creating an Elastic Workplace," *Resetting Horizons—Human Capital Trends*, Deloitte, 2013. As of July 10, 2018:
https://www2.deloitte.com/content/dam/Deloitte/global/Documents
/HumanCapital/dttl-humancapital-trends5-workplaces-no-exp.pdf

Haut Comité d'Evaluation de la Condition Militaire [Military Status Assessment Committee], *Revue Annuelle de la Condition Militaire [Annual Review of the Status of the Military]*, Paris, 2017.

Herbst, Chris M., "The Rising Cost of Child Care in the United States: A Reassessment of the Evidence," *Economics of Education Review*, Vol. 64, 2018, pp. 13–30.

Heywood, Wendy, Kent Patrick, Anthony M. A. Smith, Judy M. Simpson, Marian K. Pitts, Juliet Richters, and Julia M. Shelley, "Who Gets Tattoos? Demographic and Behavioral Correlates of Ever Being Tattooed in a Representative Sample of Men and Women," *Annals of Epidemiology*, Vol. 22, No. 1, January 2012, pp. 51–56.

Hinsliff, Gaby, "Judges, Soldiers, MPs, Vicars—Can Job-Sharing Work in Any Field?" *The Guardian*, December 2, 2016. As of October 2, 2018:
https://www.theguardian.com/money/2016/dec/02/judges-soldiers-mps-vicars-can
-job-sharing-work-in-any-field

Hodgson Russ LLP, "New U.S. DOL Memo Concludes Most Workers Are Employees, Not Independent Contractors," July 15, 2015. As of July 11, 2018:
http://www.hodgsonruss.com/newsroom-publications-7999.html

Howard, John, "Nonstandard Work Arrangements and Worker Health and Safety," *American Journal of Industrial Medicine*, Vol. 60, No. 1, January 2017, pp. 1–10.

Houseman, Susan N., *Flexible Staffing Arrangements: A Report on Temporary Help, On Call, Direct-Hire Temporary, Leased, Contract Company, and Independent Contractor Employment in the United States*, August 1999. As of May 1, 2019:
http://citeseerx.ist.psu.edu/viewdoc/download?doi=10.1.1.210.2977&rep=rep1&
type=pdf

———, "Why Employers Use Flexible Staffing Arrangements: Evidence from an Establishment Survey," *Industrial and Labor Relations Review*, Vol. 55, No. 1, October 2001, pp. 149–170.

Inavero, "2018 Future Workforce Report: Hiring Manager Insights on Flexible and Remote Work Trends," presentation slides, February 2018. As of July 17, 2018: https://www.slideshare.net/upwork/2018-future-workforce-report-hiring-manager -insights-on-flexible-and-remote-work-trends/1

Indeed.com, search for "Telemedicine Physician Work from Home." As of July 11, 2018: https://www.indeed.com/q-Telemedicine-Physician-Work-From-Home-jobs.html

Indeed Hiring Lab, *Targeting Today's Job Seeker: Data, Trends and Insight*, 2017. As of July 10, 2018: http://www.hiringlab.org/wp-content/uploads/2018/02/Hiring-Lab-Chartbook -2017_ebook-2-1-1.pdf

Internal Revenue Service, "Independent Contractor (Self-Employed) or Employee?" webpage, undated. As of July 11, 2018: https://www.irs.gov/businesses/small-businesses-self-employed/independent -contractor-self-employed-or-employee

International Labour Office, *Non-Standard Forms of Employment: Report for Discussion at the Meeting of Experts on Non-Standard Forms of Employment*, Geneva, Switzerland, February 2015. As of July 3, 2018: http://www.ilo.org/wcmsp5/groups/public/@ed_protect/@protrav/@travail /documents/meetingdocument/wcms_336934.pdf

Istrate, Emilia, and Jonathan Harris, "The Future of Work: The Rise of the Gig Economy," National Association of Counties, Counties Futures Lab, November 2017. As of July 11, 2018: http://www.naco.org/featured-resources/future-work-rise-gig-economy#ref1

Jackson, Emilie, Adam Looney, and Shanthi Ramnath, "The Rise of Alternative Work Arrangements: Evidence and Implications for Tax Filing and Benefit Coverage," Office of Tax Analysis, U.S. Department of the Treasury, Working Paper 114, January 2017. As of July 11, 2018: https://www.treasury.gov/resource-center/tax-policy/tax-analysis/Documents/WP -114.pdf

Jane's, "Estonia—Navy," July 24, 2017a. As of May 9, 2018: https://janes.ihs.com/Janes/Display/1766505

———, "Estonia—Air Force," December 29, 2017b. As of May 9, 2018: https://janes.ihs.com/Janes/Display/1319031

———, "Estonia—Army," April 11, 2018a. As of May 5, 2018: https://janes.ihs.com/Janes/Display/1319220

———, "Estonia: Executive Summary," April 12, 2018b. As of May 5, 2018: https://janes.ihs.com/Janes/Display/1302756

———, "Australia: Armed Forces," September 8, 2018c.

Joint Advertising Market Research and Studies, *Generational Values and Their Impact on Military Recruiting*, October 2016.

Joint Services Support, "Civilian Employer Information (CEI)," webpage, undated. As of February 18, 2019:
https://www.jointservicessupport.org/ESP/CEI.aspx

Jones, Charisse, "UPS and FedEx Will Hire Thousands This Holiday Season," *USA Today*, September 20, 2017. As of July 10, 2018:
https://www.usatoday.com/story/money/2017/09/20/ups-and-fedex-hire-thousands -holiday-season/686495001

Kaitseliit [Estonian Defence League], "Estonian Defence League," webpage, last updated November 10, 2018a. As of November 12, 2018:
http://www.kaitseliit.ee/en/edl

———, "Estonian Defence League Cyber Unit," webpage, last updated November 10, 2018b. As of November 12, 2018:
http://www.kaitseliit.ee/en/cyber-unit

———, "Frequently Asked Questions," webpage, last updated November 10, 2018c. As of November 12, 2018:
http://www.kaitseliit.ee/en/frequently-asked-questions

———, "The Main Tasks of the EDL CU," last updated November 10, 2018d. As of November 12, 2018:
http://www.kaitseliit.ee/en/the-main-tasks-of-the-edl-cu

———, "Members of the Estonian Defence League," webpage, last updated November 16, 2018e. As of November 16, 2018:
http://www.kaitseliit.ee/en/members

Kalleberg, Arne L., "Nonstandard Employment Relations: Part-Time, Temporary and Contract Work," *Annual Review of Sociology*, Vol. 26, 2000, pp. 341–365.

Kapp, Lawrence, and Barbara Salazar Torreon, *Reserve Component Personnel Issues: Questions and Answers*, Washington, D.C.: Congressional Research Service, January 18, 2017.

Kaska, Kadri, Anna-Maria Osula, and Jan Stinissen, *The Cyber Defence Unit of the Estonian Defence League Legal, Policy, and Organizational Analysis*, Tallinn, Estonia: NATO Cooperative Cyber Defence Center of Estonia CCDCOE, 2013. As of May 5, 2018:
https://ccdcoe.org/uploads/2018/10/CDU_Analysis.pdf

Katz, Lawrence F., and Alan B. Krueger, "The Rise and Nature of Alternative Work Arrangements in the United States, 1995–2015," National Bureau of Economic Research Working Paper No. 22667, September 2016. As of July 3, 2018:
http://www.nber.org/papers/w22667.pdf

Kelderman, Eric, "State Defense Forces Grow, Project New Image," *Stateline*, December 31, 2003. As of August 7, 2018:
https://web.archive.org/web/20120712202018/http://www.pewstates.org/projects/stateline/headlines/state-defense-forces-grow-project-new-image-85899393830

Kossek, Ellen Ernst, Leslie B. Hammer, Rebecca J. Thompson, and Lisa Buxbaum Burke, *Leveraging Workplace Flexibility for Engagement and Productivity*, Society for Human Resource Management Foundation, 2014. As of July 3, 2018:
https://www.shrm.org/hr-today/trends-and-forecasting/special-reports-and-expert-views/Documents/Leveraging-Workplace-Flexibility.pdf

Kulkarni, Nitish, "Amazon Revamps Parental Leave Policy," Techcrunch.com, November 2, 2015. As of July 11, 2018:
https://techcrunch.com/2015/11/02/amazon-revamps-parental-leave-policy/

Lambert, Susan J., Peter J. Fugiel, and Julia R. Henly, *Precarious Work Schedules Among Early-Career Employees in the US: A National Snapshot*, Chicago, Ill.: Employment Instability, Family Well-Being, and Social Policy Network, University of Chicago, August 27, 2014. As of July 10, 2018:
https://ssa.uchicago.edu/sites/default/files/uploads/lambert.fugiel.henly_.precarious_work_schedules.august2014_0.pdf

Leick, Jaime, and Kenneth Matos, *Workflex in Retail, Service, and Hospitality Guide: Cooperative Scheduling, Beyond Bias*, When Work Works, 2016. As of July 10, 2018:
http://www.whenworkworks.org/downloads/workflex-in-retail-service-hospitality-guide.pdf

———, *Workflex and Health Care Guide*, When Work Works, March 2017. As of July 10, 2018:
http://www.whenworkworks.org/downloads/workflex-and-health-care-guide.pdf

Lewis, Jennifer Lamping, Edward G. Keating, Leslie Adrienne Payne, Brian J. Gordon, Julia Pollak, Andrew Madler, H. G. Massey, and Gillian S. Oak, *U.S. Department of Defense Experiences with Substituting Government Employees for Military Personnel: Challenges and Opportunities*, Santa Monica, Calif.: RAND Corporation, RR-1282-OSD, 2016. As of July 19, 2018:
https://www.rand.org/pubs/research_reports/RR1282.html

Lincoln, Paul, foreword to *A History of Our Reserves*, UK Ministry of Defence, 2014. As of May 2, 2018:
https://assets.publishing.service.gov.uk/government/uploads/system/uploads/attachment_data/file/283812/a-history-our-Reserves-Epub-v2.pdf

Maestas, Nicole, Kathleen J. Mullen, David Powell, Till von Wachter, and Jeffrey B. Wenger, *The American Working Conditions Survey Data: Codebook and Data Description*, Santa Monica, Calif.: RAND Corporation, TL-269-APSF, 2017. As of May 1, 2019:
https://www.rand.org/pubs/tools/TL269.html

Marine Corps Order P1100.72C, *Military Personnel Procurement Manual, Volume 2: Enlisted Procurement*, June 18, 2004. As of September 11, 2018: https://www.marines.mil/Portals/59/Publications/MCO%20P1100.72C%20W%20ERRATUM.pdf

Maryland Defense Force, homepage, undated(a). As of November 12, 2018: http://mddf.maryland.gov

———, "MDDF Units," webpage, undated(b). As of August 8, 2018: http://military.maryland.gov/mddf/Pages/MDDF-Units.aspx

Mas, Alexandre, and Amanda Pallais, "Valuing Alternative Work Arrangements," March 2017. As of July 10, 2018: https://scholar.harvard.edu/files/pallais/files/alternative_work_arrangements.pdf

Mastracci, Sharon H., and James R. Thompson, "Nonstandard Work Arrangements in the Public Sector: Trends and Issues," *Review of Public Personnel Administration*, Vol. 25, No. 4, December 2005, pp. 299–324.

———, "Who Are the Contingent Workers in Federal Government," *American Review of Public Administration*, Vol. 39, No. 4, July 2009, pp. 352–373.

Matos, Kenneth, *Workflex and Telework Guide: Everyone's Guide to Working Anywhere*, When Work Works, 2015. As of July 10, 2018: http://www.whenworkworks.org/downloads/workflex-and-telework-guide.pdf

Matos, Kenneth, Ellen Galinsky, and James T. Bond, *2016 National Study of Employers*, Society for Human Resources Management, Families and Work Institute, and When Work Works, 2017. As of July 10, 2018: http://whenworkworks.org/downloads/2016-National-Study-of-Employers.pdf

Matos, Kenneth, and Eve Tahmincioglu, *Workflex and Manufacturing Guide: More Than a Dream*, When Work Works, 2015. As of July 10, 2018: http://www.whenworkworks.org/downloads/manufacturing-guide.pdf

McKay, Gretchen, "Illustrated Chefs: Why Kitchen Artists Are Big on Tattoos," *Pittsburgh Post-Gazette*, June 19, 2011.

McKinsey Global Institute, *Independent Work: Choice, Necessity, and the Gig Economy*, October 2016. As of July 11, 2018: https://www.mckinsey.com/~/media/mckinsey/featured%20insights/Employment%20and%20Growth/Independent%20work%20Choice%20necessity%20and%20the%20gig%20economy/Independent-Work-Choice-necessity-and-the-gig-economy-Full-report.ashx

Ministère des Armées [French Ministry of the Armed Forces], "Réserviste au Service de santé des armées (SSA): une vie pleine d'aventure," La réserve militaire, composante de la garde nationale, July 18, 2016. As of May 2, 2019: https://www.defense.gouv.fr/reserve/actualites/reserviste-au-service-de-sante-des-armees-ssa-une-vie-pleine-d-aventure

————, *Rapport d'évaluation de la réserve militaire 2016* [*Military Reserve Evaluation Report 2016*], Paris, 2017.

————, "La Garde Nationale," undated(a). As of May 2, 2019: https://www
.defense.gouv.fr/content/download/508360/8594841/file/2017_d%C3%A9pliant
_GardeNat_pilier%20MINDEF.pdf

Moore, Nancy Young, Molly Dunigan, Frank Camm, Samantha Cherney, Clifford A. Grammich, Judith D. Mele, Evan Peet, and Anita Szafran, *A Review of Alternative Methods to Inventory Contracted Services in the Department of Defense*, Santa Monica, Calif.: RAND Corporation, RR-1704-OSD, 2017. As of July 19, 2018:
https://www.rand.org/pubs/research_reports/RR1704.html

Morin, Rich, "For Many Injured Veterans, a Lifetime of Consequences," Pew Research Center, November 8, 2011. As of October 2, 2018:
http://www.pewsocialtrends.org/2011/11/08/for-many-injured-veterans-a-lifetime
-of-consequences

Mysliwiec, Vincent, Leigh McGraw, Roslyn Pierce, Patrick Smith, Brandon Trapp, and Bernard J. Roth, "Sleep Disorders and Associated Medical Comorbidities in Active Duty Military Personnel," *Sleep*, Vol. 36, No. 2, 2013, pp. 167–174.

National Academies of Sciences, Engineering, and Medicine, *Information Technology and the U.S. Workforce: Where Are We and Where Do We Go from Here?*, Washington, D.C.: The National Academies Press, 2017. As of March 19, 2019: https://www.nap.edu/catalog/24649/information-technology-and-the-us
-workforce-where-are-we-and

National Archives and Records Administration, "Request for Records Disposition Authority," Worldwide Individual Augmentation System, Records Schedule No. DAA-AU-2016-0036, September 20, 2016. As of August 9, 2018:
https://www.archives.gov/files/records-mgmt/rcs/schedules/departments
/department-of-defense/department-of-the-army/rg-au/daa-au-2016-0036_sf115
.pdf

National Archives of Australia, "Universal Military Training in Australia, 1911–29," Fact Sheet No. 160, undated. As of November 12, 2018:
http://www.naa.gov.au/collection/fact-sheets/fs160.aspx

National Interagency Fire Center, "National Interagency Incident Management Team Rotation," November 2018. As of July 19, 2017:
https://www.nifc.gov/nicc/logistics/teams/imt_rotate.pdf

National Opinion Research Center, *General Social Surveys—Quality of Working Life Module, 1972–2014: Cumulative Codebook*, Chicago, Ill.: University of Chicago, June 2017. As of October 2, 2018:
http://gss.norc.org/Documents/codebook/QWL%20Codebook.pdf

National Organization on Disability, *Return to Careers*, New York, November 2011. As of August 7, 2018:
https://www.nod.org/wp-content/uploads/Return-to-Careers-Full-Report_rev2018
.pdf

National Wildfire Coordinating Group, *NWCG Standards for Interagency Incident Business Management*, Boise, Id., PMS 902, April 2018. As of November 12, 2018:
https://www.nwcg.gov/sites/default/files/publications/pms902.pdf

Naval Reserve Officers Training Corps, "Strategic Sealift Midshipman Program (SSMP)," webpage, undated. As of February 18, 2019:
https://www.nrotc.navy.mil/ssmp.html

Navy CyberSpace, "Navy Dependency Waiver," webpage, June 17, 2018. As of October 2, 2018:
https://www.navycs.com/blogs/2011/01/07/navy-dependency-waiver

Nguyen, Steve, "Results-Only Work Environment (ROWE)," *Workplace Psychology*, January 4, 2017. As of November 12, 2018:
https://workplacepsychology.net/2017/01/04/results-only-work-environment-rowe/

Nicholson, Jessica R., "Temporary Help Workers in the U.S. Labor Market," Washington, D.C.: Economics and Statistics Administration, U.S. Department of Commerce, Issue Brief 03-15, July 1, 2015.

NORC—*See* National Opinion Research Center.

Office of Management and Budget, "Policy Letter 11-01, Performance of Inherently Governmental and Critical Functions," *Federal Register*, Vol. 76, No. 176, September 12, 2011, pp. 56227–56242. As of July 19, 2018:
https://www.gpo.gov/fdsys/pkg/FR-2011-09-12/pdf/2011-23165.pdf

———, "Policy Letter 11–01, Performance of Inherently Governmental and Critical Functions," *Federal Register*, Vol. 77, No. 29, February 13, 2012, p. 7609. As of November 12, 2018:
https://www.federalregister.gov/documents/2012/02/13/2012-3190/policy-letter-11
-01-performance-of-inherently-governmental-and-critical-functions

Office of Personnel Management, *Compensation Flexibilities to Recruit and Retain Cybersecurity Professionals*, Washington, D.C., undated(a). As of July 17, 2018:
https://www.opm.gov/policy-data-oversight/pay-leave/reference-materials
/handbooks/compensation-flexibilities-to-recruit-and-retain-cybersecurity
-professionals.pdf

———, *Handbook on Alternative Work Schedules*, Washington, D.C., undated(b). As of July 17, 2018:
https://www.opm.gov/policy-data-oversight/pay-leave/reference-materials
/handbooks/alternative-work-schedules/

————, "Veterans Services," webpage, undated(c). As of October 2, 2018: https://www.opm.gov/policy-data-oversight/veterans-services/vet-guide-for-hr -professionals

————, *Human Resources Flexibilities and Authorities in the Federal Government*, Washington, D.C., August 2013. As of July 17, 2018: https://www.opm.gov/policy-data-oversight/pay-leave/reference-materials /handbooks/humanresourcesflexibilitiesauthorities.pdf

————, *Status of Telework in the Federal Government: Report to Congress Fiscal Year 2016*, Washington, D.C., November 2017. As of July 17, 2018: https://www.telework.gov/reports-studies/reports-to-congress/2017-report-to -congress.pdf

————, "Employment Cubes," FedScope database, last updated March 2018. As of July 19, 2018: https://www.fedscope.opm.gov/employment.asp

Office of the Under Secretary of Defense for Personnel and Readiness, *Population Representation in the Military Services: Fiscal Year 2014 Summary Report*, Washington, D.C., March 1, 2016.

————, "DoD Total Military Strength: January FY2019," undated, document provided to the authors, February 22, 2019.

OPM—*See* Office of Personnel Management.

Organisation for Economic Co-operation and Development, "LMF2.4. Family-Friendly Workplace Policies," Organisation for Economic Co-operation and Development Family Database, December 8, 2016. As of July 3, 2018: https://www.oecd.org/els/family/LMF_2-4-Family-friendly-workplace-practices .pdf

————, *Be Flexible! Background Brief on How Workplace Flexibility Can Help European Employees to Balance Work and Family*, September 2016. As of July 11, 2018: https://www.oecd.org/els/family/Be-Flexible-Backgrounder-Workplace-Flexibility .pdf

Orvis, Bruce R., Herb Shukiar, Laurie L. McDonald, Michael G. Mattock, M. Rebecca Kilburn, and Michael G. Shanley, *Ensuring Personnel Readiness in the Army Reserve Components*, Santa Monica, Calif.: RAND Corporation, MR-659-A, 1996. As of February 25, 2019: https://www.rand.org/pubs/monograph_reports /MR659.html

OSD—*See* Office of the Secretary of Defense.

Pack, Al, and G. W. Pien, "Update on Sleep and Its Disorders," *Annual Review of Medicine*, Vol. 62, 2011, pp. 447–460.

Padar, Andrus, Commander, Cyber Defense Unit, "Estonian Defence League Cyber Defence Unit," presentation, July 2017.

Peterson, Michael T., *The Reinvention of the Canadian Armed Forces Chaplaincy and the Limits of Religious Pluralism*, thesis, Wilfrid Laurier University, 2015. As of November 8, 2018:
http://scholars.wlu.ca/etd/1729

Pew Research Center, *Parenting in America: Outlook, Worries, Aspirations Are Strongly Linked to Financial Situation*, Washington, D.C., December 17, 2015. As of November 8, 2018:
http://www.pewresearch.org/wp-content/uploads/sites/3/2015/12/2015-12-17 _parenting-in-america_FINAL.pdf

Pfeffer, Jeffrey, and James N. Baron, "Taking the Workers Back Out: Recent Trends in the Structuring of Employment," in Barry M. Staw and Larry L. Cummings, eds., *Research in Organizational Behavior*, Vol. 10, Greenwich, Conn.: JAI Press, 1988, pp. 257–303.

Polivka, Anne E., "Contingent and Alternative Work Arrangements, Defined," *Monthly Labor Review*, Bureau of Labor Statistics, October 1996. As of July 3, 2018:
https://www.bls.gov/opub/mlr/1996/10/art1full.pdf

Poncelin de Raucourt, Gaëtan, *Compte rendu: Commission de la défense nationale et des forces armées [Report: National Defense and Armed Forces Commission]*, Paris: French National Assembly, 2017.

Public Law 109–163, National Defense Authorization Act for Fiscal Year 2006, Sec. 514, January 6, 2006.

Public Law 111-292, Telework Enhancement Act of 2010, December 9, 2010.

Punaro, Arnold L., Chair, Reserve Forces Policy Board, "Report of the Reserve Forces Policy Board on the 'Operational Reserve' and Inclusion of the Reserve Components in Key Department of Defense (DoD) Processes," memorandum to the Secretary of Defense, January 14, 2013a.

———, "Report of the Reserve Forces Policy Board on Reserve Component (RC) Duty Status Reform," memorandum, July 16, 2013b. As of October 2, 2018:
https://rfpb.defense.gov/Portals/6//Documents/RFPB_Duty_Status_rpt_to _SECDEF.pdf

Purtill, Corrine, "PWC's Millennial Employees Led a Rebellion—and Their Demands Are Being Met," *Quartz at Work*, March 20, 2018. As of July 12, 2018:
https://work.qz.com/1217854/pwcs-millennial-employees-led-a-rebellion-and-their -demands-are-being-met

PwC, "PwC's Flexibility² Talent Network—the Best of Both Worlds," webpage, undated(a). As of July 12, 2018:
https://www.pwc.com/us/en/careers/why-pwc/flexibility-talent-network.html

———, "PwC's Flexibility² Talent Network FAQs," webpage, undated(b). As of July 12, 2018:
https://www.pwc.com/us/en/careers/why-pwc/ftn_faqs.html

———, "Quality of life: Balancing Work-Life Demands," webpage, undated(c). As of July 12, 2018:
https://www.pwc.com/us/en/about-us/diversity/pwc-work-life-balance.html

———, "Flexibility² Talent Network," fact sheet, 2015. As of July 12, 2018:
https://www.pwc.com/us/en/careers/experienced/why-pwc/links/ftn-sell-sheet-final
.pdf

Rajaratnam, Shantha, Laura K. Barger, Steven W. Lockley, Steven A. Shea, Wei Wang, Christopher P. Landrigan, Conor S. O'Brien, Salim Qadri, Jason P. Sullivan, Brian E. Cade, Lawrence J. Epstein, David P. White, and Charles A. Czeisler, "Sleep Disorders, Health and Safety in Police Officers," *Journal of the American Medical Association*, Vol. 306, No. 23, 2011, pp. 2567–2578.

Randall, Scott, "Can I Drill from Home? Telework (or the Lack Thereof) in the Army Reserve," *Army Lawyer*, April 2014. As of July 19, 2018:
https://tjaglcspublic.army.mil/documents/27431/2101332/2014-Apr-Randall-Drill%20and%20Telework%20in%20the%20Army%20Reserve.pdf/4940f02c
-1e32-41c9-a2cc-006cb17c2869

RediDoc, homepage, undated. As of July 11, 2018:
http://redidoc.com

Republic of Estonia, Ministry of Defence, "Defence and Armed Forces," webpage, last updated March 15, 2016. As of May 5, 2018:
http://www.kmin.ee/en/objectives-activities/defence-and-armed-forces

Reserve Forces Policy Board, *Eliminating Major Gaps in DoD Data on the Fully-Burdened and Life-Cycle Cost of Military Personnel: Cost Elements Should Be Mandated by Policy*, Final Report to the Secretary of Defense (RFPB Report FY13-02), Washington, D.C., January 7, 2013.

Reynolds, Brie Weiler, "5 of the Fastest-Growing Remote Career Categories," Flexjobs.com, January 3, 2018. As of July 19, 2018:
https://www.flexjobs.com/blog/post/fastest-growing-remote-career-categories/

Robles, Barbara, and Marysol McGee, "Exploring Online and Offline Informal Work: Findings from the Enterprising and Informal Work Activities (EIWA) Survey," Finance and Economics Discussion Series 2016-089, Washington, D.C.: Board of Governors of the Federal Reserve System, 2016. As of July 3, 2018:
https://doi.org/10.17016/FEDS.2016.089

Ruiz, Monica M., "Is Estonia's Approach to Cyber Defense Feasible in the United States?" *War on the Rocks*, January 9, 2018. As of April 29, 2018:
https://warontherocks.com/2018/01/estonias-approach-cyber-defense-feasible
-united-states

Shambaugh, Jay, Ryan Nunn, and Lauren Bauer, "Independent Workers and the Modern Labor Market," Brookings Institution, June 7, 2018. As of February 18, 2019:
https://www.brookings.edu/blog/up-front/2018/06/07/independent-workers-and -the-modern-labor-market

Shin, Doosup, Chandrashekar Bohra, Kullatham Kongpakpaisarn, and Eun Sun Lee, "Increasing Trend in the Prevalence of Abdominal Obesity in the United States During 2001–2016," *Journal of the American College of Cardiology*, Vol. 71, No. 11, supplement, March 2018, p. A1737.

Shin, Laura, "Work from Home 2018: The Top 100 Companies for Remote Jobs," *Forbes*, January 17, 2018. As of August 9, 2018:
https://www.forbes.com/sites/laurashin/2018/01/17/work-from-home-2018-the-top -100-companies-for-remote-jobs

Sloan Center on Aging and Work, "CVS Caremark Snowbird Program," Innovative Practices Database, 2012. As of July 11, 2018:
http://capricorn.bc.edu/agingandwork/database81/browse/case_study/24047/

Small Business Administration, "Office of Disaster Assistance Employment," webpage, undated. As of July 19, 2018:
https://www.sba.gov/loans-grants/see-what-sba-offers/sba-loan-programs/disaster -loans/office-disaster-assistance-employment

———, "Office of Disaster Assistance Staffing Strategy," Washington, D.C., June 4, 2014.

Smith, Hugh, and Nick Jans, "Use Them or Lose Them? Australia's Defence Force Reserves," *Armed Forces and Society*, Vol. 37, No. 2, 2011, pp. 301–320.

Statista, "Do You Have a Tattoo? (by Occupation)," 2013 data, webpage, undated. As of October 2, 2018:
https://www.statista.com/statistics/259588/share-of-americans-with-a-tattoo-by -occupation

Tamkin, Emily, "10 Years After the Landmark Attack on Estonia, Is the World Better Prepared for Cyber Threats?" *Foreign Policy*, April 27, 2017. As of November 12, 2018:
https://foreignpolicy.com/2017/04/27/10-years-after-the-landmark-attack-on -estonia-is-the-world-better-prepared-for-cyber-threats/

Taylor, Paul, *For Many Injured Veterans, a Lifetime of Consequences*, Washington, D.C.: Pew Social and Demographic Trends, November 8, 2011.

Taylor, Paul, and Scott Keeter, eds., *Millennials: A Portrait of Generation Next*, Washington, D.C.: Pew Research Center, February 2010. As of November 8, 2018:
http://www.pewresearch.org/wp-content/uploads/sites/3/2010/10/millennials -confident-connected-open-to-change.pdf

Telework.gov, "Telework Legislation," webpage, OPM, undated(a). As of July 17, 2018:
https://www.telework.gov/guidance-legislation/telework-legislation/legislation/

―――, "Telework Managers: Assessing Job Tasks," webpage, OPM, undated(b). As of July 10, 2018:
https://www.telework.gov/federal-community/telework-managers/assessing-job-tasks

Torpey, Elka Maria, "Flexible Work: Adjusting the Where and When of Your Job," *Occupational Outlook Quarterly*, Summer 2017, pp. 14–27. As of July 3, 2018:
https://www.bls.gov/careeroutlook/2007/summer/art02.pdf

Treverton, Gregory F., David M. Oaks, Lynn Scott, Justin L. Adams, and Stephen Dalzell, *Attracting "Cutting-Edge" Skills Through Reserve Component Participation*, Santa Monica, Calif.: RAND Corporation, MR-1729-OSD, 2003. As of February 13, 2019: https://www.rand.org/pubs/monograph_reports/MR1729.html

TRICARE, "Covered Services: Telemedicine," webpage, undated(a). As of August 9, 2018:
https://tricare.mil/CoveredServices/IsItCovered/Telemedicine

―――, "TRICARE Reserve Select," webpage, undated(b). As of October 2, 2018:
https://tricare.mil/TRS

Turner, Paul, "My Two Months of Seasonal Work at an Amazon Fulfillment Center," *Billfold*, February 4, 2016. As of July 10, 2018:
https://www.thebillfold.com/2016/02/my-two-months-of-seasonal-work-at-an-amazon-fulfillment-center

Under Secretary of Defense for Personnel and Readiness, *Defense Manpower Requirements Report, Fiscal Year 2003*, April 2002. As of August 10, 2018:
https://prhome.defense.gov/Portals/52/Documents/RFM/TFPRQ/Docs/FY2003.pdf

Union for Reform Judaism, "Support for Jewish Military Chaplains and Jewish Military Personnel and Their Families," resolution, 2005. As of November 8, 2018:
https://urj.org/what-we-believe/resolutions/support-jewish-military-chaplains-and-jewish-military-personnel-and

UK Government, "Working for JFC: A Joint Cyber Reserve Force," webpage, undated. As of May 2, 2018:
https://www.gov.uk/government/organisations/joint-forces-command/about/recruitment

UK Ministry of Defence, *A History of Our Reserves*, London, 2014. As of May 3, 2018:
https://assets.publishing.service.gov.uk/government/uploads/system/uploads/attachment_data/file/283812/a-history-our-Reserves-Epub-v2.pdf

———, *UK Reserve Forces and Cadets Strengths, 1 April 2015*, London, June 18, 2015. As of May 2, 2018:
https://assets.publishing.service.gov.uk/government/uploads/system/uploads/attachment_data/file/435444/20150615_TSP7Apr15-O.pdf

———, "UK Armed Forces Quarterly Service Personnel Statistics, 1 July 2018," August 23, 2018. As of November 12, 2018:
https://www.gov.uk/government/statistics/quarterly-service-personnel-statistics-2018

U.S. Air Force Auxiliary, Civil Air Patrol, "Who We Are," webpage, undated. As of October 17, 2018:
https://www.gocivilairpatrol.com/about/who-we-are

U.S. Air Force Personnel Center, "Selective Retention Bonus Listing," May 30, 2018. As of October 2, 2018:
http://www.afpc.af.mil/Portals/70/documents/14_RETENTION/FY18%20Mid-Year%20SRB%20List.pdf?ver=2018-05-29-132042-840

U.S. Air Force Reserve, "About," webpage, undated. As of October 2, 2018:
https://afreserve.com/about

U.S. Army, "Army Civilian Acquired Skills Program ACASP," last modified March 8, 2011. As of May 2, 2019:
http://www.army-portal.com/jobs/acasp.html

U.S. Army, "Benefits," webpage, undated. As of July 2, 2018:
https://www.goarmy.com/benefits/money/bonuses-earning-extra-money.html

U.S. Army Human Resources Command, "Volunteering for Worldwide Individual Augmentation System (WIAS) Positions," webpage, last updated May 14, 2018. As of August 9, 2018:
https://www.hrc.army.mil/content/HOT%20TOPICS%20-%20Enlisted%20Personnel%20Management%20Directorate

U.S. Army Recruiting Regulation 6601-566, *Personnel Procurement: Waiver, Future Soldier Program Separation, and Void Enlistment Processing Procedures*, May 31, 2006.

U.S. Army Reserve, "Ways to Serve: Troop Program Units (TPUs)," webpage, undated. As of November 8, 2018:
https://www.usar.army.mil/TPU

U.S.C.—*See* U.S. Code.

U.S. Census Bureau, "American Community Survey (ACS)," webpage, undated. As of November 12, 2018:
https://www.census.gov/programs-surveys/acs/

———, "Mother's Day: May 8, 2016," press release No. CB16-FF.09, April 20, 2016. As of November 8, 2018:
https://www.census.gov/newsroom/facts-for-features/2016/cb16-ff09.html

U.S. Coast Guard Auxiliary, "About the Auxiliary," webpage, undated. As of August 8, 2018:
http://cgaux.org/about.php

U.S. Code, Title 5, Government Organizations and Employees, Section 3109, Employment of Experts and Consultants; Temporary or Intermittent, January 5, 1999.

U.S. Code, Title 10, Armed Forces, Section 129a, General Policy for Total Force Management, January 12, 2018.

U.S. Code, Title 10, Armed Forces, Section 129b, Authority to Procure Personal Services, January 12, 2018.

U.S. Code, Title 10, Armed Forces, Section 10102, Purpose of Reserve Components, January 12, 2018.

U.S. Code, Title 10, Armed Forces, Section 10143, Ready Reserve: Selected Reserve, January 12, 2018.

U.S. Code, Title 10, Armed Forces, Section 10144, Ready Reserve: Individual Ready Reserve, January 12, 2018.

U.S. Code, Title 10, Armed Forces, Section 10147, Ready Reserve: Training Requirements, January 12, 2018.

U.S. Code, Title 10, Armed Forces, Section 10216, Military Technicians (Dual Status), January 12, 2018.

U.S. Code, Title 10, Armed Forces, Section 12304, Selected Reserve and Certain Individual Ready Reserve Members; Order to Active Duty Other Than During War or National Emergency, January 12, 2018.

U.S. Code, Title 10, Armed Forces, Section 12310, Reserves: For Organizing, Administering, etc., Reserve Components, January 12, 2018.

U.S. Code, Title 38, Veterans' Benefits, Section 8127, Small Business Concerns Owned and Controlled by Veterans: Contracting Goals and Preferences, January 3, 2012.

U.S. Code, Title 50, War and National Defense, Section 3901, Servicemembers Civil Relief Act, January 12, 2018.

U.S. Department of Defense, "Military Compensation," webpage, undated. As of October 2, 2018:
https://militarypay.defense.gov/Pay/Retirement/Reserve.aspx

———, *Quadrennial Defense Review 2014*, March 4, 2014. As of November 8, 2018:
http://archive.defense.gov/pubs/2014_quadrennial_defense_review.pdf

———, *2016 Demographics: Profile of the Military Community*, Washington, D.C., 2016. As of October 2, 2018:
http://download.militaryonesource.mil/12038/MOS/Reports/2016-Demographics -Report.pdf

———, *Enhancement of Use of Telehealth Services in the Military Health System*, report in response to Section 718 of the National Defense Authorization Act for Fiscal Year 2017 (Public Law 114-328), October 2017. As of August 9, 2018:
https://health.mil/Reference-Center/Reports/2017/10/17/Enhancement-of-Use-of -Telehealth-Services-in-the-MHS

U.S. Department of Defense, Civilian Personnel Advisory Service, "Cyber Hiring Options/Authorities Guide," December 2016. As of July 17, 2018:
https://www.cpms.osd.mil/Content/Documents/CyberHiringOptions _AuthoritiesGuidev6.pdf

U.S. Department of Defense, Office of the Actuary, *Statistical Report on the Military Retirement System, Fiscal Year 2016*, July, 2017.

U.S. Department of Defense Directive 1100.4, *Guidance for Manpower Management*, Washington, D.C., February 12, 2005.

U.S. Department of Defense Instruction 1035.10, *Telework Policy*, Washington, D.C., April 4, 2012.

———, 1100.22, *Policy and Procedures for Determining Workforce Mix*, Washington, D.C., incorporating change 1, December 1, 2017.

———, 1215.06, *Uniform Reserve, Training, and Retirement Categories for the Reserve Components*, Washington, D.C., incorporating change 1, May 19, 2015.

———, 1205.18, *Full-Time Support to the Reserve Components*, Washington, D.C., May 12, 2014.

———, 1300.17, *Accommodation of Religious Practices Within the Military Services*, Washington, D.C., incorporating change 1, January 22, 2014.

———, 1304.26, *Qualification Standards for Enlistment, Appointment, and Induction*, Washington, D.C., incorporating change 2, April 11, 2017.

U.S. Department of Labor, "US Secretary of Labor Withdraws Joint Employment, Independent Contractor Informal Guidance," press release, June 7, 2017. As of July 11, 2018:
https://www.dol.gov/newsroom/releases/opa/opa20170607

U.S. Department of the Navy, *Fiscal Year 2019 Budget Estimates, Justification of Estimates*, Washington, D.C., February 2018.

U.S. Department of Veterans Affairs, *Veterans Benefits Administration Annual Benefits Report, Fiscal Year 2016*, Washington, D.C., 2017. As of November 12, 2018:
https://www.benefits.va.gov/REPORTS/abr/docs/2016_abr.pdf

U.S. General Services Administration, Federal Acquisition Regulation Subpart 19.14, Service-Disabled Veteran-Owned Small Business Procurement Program.

———, "Tools for Effective Telework," webpage, undated. As of July 11, 2018:
https://www.gsa.gov/governmentwide-initiatives/telework/tools-for-effective-telework

U.S. Government Accountability Office, "Contingent Workforce: Size, Characteristics, Earnings, and Benefits," memorandum to Sen. Patty Murray, ranking member, U.S. Senate Committee on Health, Education, Labor, and Pensions, and Sen. Kirsten Gillibrand, GAO-15-168R, April 20, 2015.

———, *DOD Inventory of Contracted Services: Timely Decisions and Further Actions Needed to Address Long-Standing Issues*, Washington, D.C., GAO-17-17, October 2016.

———, *Federal Contracting: Improvements Needed in How Some Agencies Report Personal Services Contracts*, Washington, D.C., GAO-17-610, July 2017a. As of July 19, 2018: https://www.gao.gov/assets/690/686179.pdf

———, *Reserve Component Travel: DOD Should Assess the Effect of Reservists' Unreimbursed Out-of-Pocket Expenses on Retention*, Washington, D.C., GAO-18-181, October 2017b.

———, "Department of Defense: Telehealth Use in Fiscal Year 2016," memorandum to congressional committees, GAO-18-108R, November 14, 2017c.

———, *Military Personnel: Additional Actions Needed to Address Gaps in Military Physician Specialties*, Washington, D.C., GAO-18-77, February 2018a.

———, *Military Personnel: DOD Needs to Reevaluate Fighter Pilot Workforce Requirements*, Washington, D.C., GAO-18-113, April 11, 2018b.

———, *Military Personnel: Collecting Additional Data Could Enhance Pilot Retention Efforts*, Washington, D.C., GAO-18-439, June 2018c.

U.S. Navy, "Navy Reservist Roles and Responsibilities," webpage, undated. As of October 2, 2018:
https://www.navy.com/about/about-reserve/structure.html

USFS—*See* U.S. Forest Service.

U.S. Forest Service, *Interagency Incident Business Management Handbook*, FSH 5109.34, April 1, 2017.

VA—*See* U.S. Department of Veterans Affairs.

Vincent, Faustine, "La garde nationale est un simple label," *Le Monde*, October 12, 2016. As of May 2, 2019:
https://www.lemonde.fr/societe/article/2016/10/12/la-garde-nationale-est-un
-simple-label_5012604_3224.html

Wagner, Alex, and Andrew Burt, "Blurred Lines: An Argument for a More Robust Legal Framework Governing the CIA Drone Program," *Yale Journal of International Law Online*, Vol. 38, Fall 2012. As of November 12, 2018:
http://www.yjil.yale.edu/volume-38

Warner, Senator Mark R., "Senator Warner Addresses the Opportunities and Challenges of the 'Sharing Economy,'" webpage, undated. As of July 11, 2018:
https://www.warner.senate.gov/public/index.cfm/on-demand-economy

Wartzman, Rick, "What America's Aging Workers Mean for the Future of Work," *Fortune*, June 22, 2016. As of July 11, 2018:
http://fortune.com/2016/06/22/aging-workforce-retirement/

Watson, Liz, and Jennifer E. Swanberg, *Flexible Workplace Solutions for Low-Wage Hourly Workers: A Framework for a National Conversation*, Workplace Flexibility 2010 at Georgetown Law School and Institute for Workplace Innovation at the University of Kentucky, May 2011. As of July 10, 2018:
http://workplaceflexibility2010.org/images/uploads/whatsnew/Flexible
%20Workplace%20Solutions%20for%20Low-Wage%20Hourly%20Workers.pdf

Weil, David, "Lots of Employees Get Misclassified as Contractors: Here's Why IT Matters," *Harvard Business Review*, July 5, 2017. As of May 1, 2019:
https://hbr.org/2017/07/lots-of-employees-get-misclassified-as-contractors-heres
-why-it-matters

Weil, David, and Tanya Goldman, "Labor Standards, the Fissured Workplace, and the On-Demand Economy," *Perspectives on Work*, 2016, pp. 26–77.

Weinbaum, Cortney, Arthur Chan, Karlyn D. Stanley, and Abby Schendt, *Moving to the Unclassified: How the Intelligence Community Can Work from Unclassified Facilities*, Santa Monica, Calif.: RAND Corporation, RR-2024-OSD, 2018. As of July 17, 2018:
https://www.rand.org/pubs/research_reports/RR2024.html

Weinbaum, Cortney, Bonnie L. Triezenberg, Erika Meza, and David Luckey, *Understanding Government Telework: An Examination of Research Literature and Practices from Government Agencies*, Santa Monica, Calif.: RAND Corporation, RR-2023-OSD, 2018. As of July 17, 2018:
https://www.rand.org/pubs/research_reports/RR2023.html

Weisberg, Ann, Deloitte, "Mass Career Customization: Building the Corporate Lattice Organization," presentation at Work/Life Policy, Practice and Potential, UN Expert Group Meeting, November 9, 2010. As of July 11, 2018:
http://www.un.org/womenwatch/osagi/pdf/FPW/worklife/The-Career-Lattice-and
-Mass-Career-Customization,-Anne-Weisberg.pdf

Wenger, Jennie W., Caolionn O'Connell, and Linda Cottrell, *Examination of Recent Deployment Experience Across the Services and Components*, Santa Monica, Calif.: RAND Corporation, RR-1928-A, 2018. As of August 14, 2018:
https://www.rand.org/pubs/research_reports/RR1928.html

When Work Works, "Workplace Awards," webpage, undated. As of July 19, 2018:
http://www.whenworkworks.org/search-recipients

Whitley, John E., James M. Bishop, Sarah K. Burns, Kristen M. Guerrera, Philip M. Lurie, Brian Q. Rieksts, Bryan W. Roberts, Timothy J. Wojtecki, and Linda Wu, *Medical Total Force Management: Assessing Readiness and Cost, Institute for Defense Analysis*, Washington, D.C.: Institute for Defense Analyses, Paper P-8805, May 2018. As of November 9, 2018:
https://prhome.defense.gov/Portals/52/Documents/MRA_Docs/TFM/Reports
/IDA%20Paper%20P-8805_Final_051518.pdf?ver=2018-06-26-105159-957

Wilke, Robert L., Under Secretary of Defense for Personnel and Readiness, "DoD Retention Policy for Non-Deployable Service Members," memorandum, February 14, 2018. As of October 2, 2018:
https://dod.defense.gov/Portals/1/Documents/pubs/DoD-Universal-Retention
-Policy.pdf

Winfield, Nicole, "Vatican Seeks 'Courageous' Ideas to Combat Priest Shortage," Associated Press, June 8, 2018. As of November 8, 2018:
https://www.apnews.com/32a3798435694ff58924e9c8c8b006da

Winston, Kimberly, "Defense Department Expands Its List of Recognized Religions," Religion News Service, April 21, 2017. As of August 8, 2018:
https://religionnews.com/2017/04/21/defense-department-expands-its-list-of
-recognized-religions

Winter, Steve, "Two Years in, Our Parental Leave Policy Is Working for Parents," *dayone: The Amazon Blog*, August 21, 2017. As of July 11, 2018:
https://blog.aboutamazon.com/working-at-amazon/two-years-in-our-parental
-leave-policy-is-working-for-all-employees

Workplace Flexibility 2010 at Georgetown University Law Center, "Flexible Work Arrangements: A Definition and Examples," undated. As of July 3, 2018:
http://workplaceflexibility2010.org/images/uploads/general_information/fwa
_definitionsexamples.pdf

Younger, Jon, and Norm Smallwood, "Performance Management in the Gig Economy," *Harvard Business Review*, January 11, 2016. As of November 7, 2018:
https://hbr.org/2016/01/performance-management-in-the-gig-economy